String Theory
Demystified

Demystified Series

String Theory
Demystified

David McMahon

New York   Chicago   San Francisco   Lisbon   London
Madrid   Mexico City   Milan   New Delhi   San Juan
Seoul   Singapore   Sydney   Toronto

Copyright © 2009 by The McGraw-Hill Companies, Inc. All rights reserved. Printed in the United States of America. Except as permitted under the United States Copyright Act of 1976, no part of this publication may be reproduced or distributed in any form or by any means, or stored in a data base or retrieval system, without the prior written permission of the publisher.

1 2 3 4 5 6 7 8 9 0   DOC/DOC   0 1 4 3 2 1 0 9 8

ISBN 978-0-07-149870-8
MHID 0-07-149870-2

Sponsoring Editor
Judy Bass

Production Supervisor
Pamela A. Pelton

Editing Supervisor
Stephen M. Smith

Project Manager
Harleen Chopra, International Typesetting and Composition

Copy Editors
Priyanka Sinha and Surendra Nath, International Typesetting and Composition

Proofreaders
Ragini Pandey and Megha RC, International Typesetting and Composition

Indexer
Broccoli Information Management

Art Director, Cover
Jeff Weeks

Composition
International Typesetting and Composition

Printed and bound by RR Donnelley.

McGraw-Hill books are available at special quantity discounts to use as premiums and sales promotions, or for use in corporate training programs. To contact a special sales representative, please visit the Contact Us page at www.mhprofessional.com.

CONTENTS

Contents

ABOUT THE AUTHOR

David McMahon has worked for several years as a physicist and researcher at Sandia National Laboratories. He is the author of *Linear Algebra Demystified, Quantum Mechanics Demystified, Relativity Demystified, MATLAB® Demystified*, and *Complex Variables Demystified*, among other successful titles.

PREFACE

String theory is the greatest scientific quest of all time. Its goal is nothing other than a complete description of physical reality—at least at the level of fundamental particles, interactions, and perhaps space-time itself. In principle, once the fundamental theory is fully known, one could derive relativity and quantum theory as low-energy limits to strings. The theory sets out to do what no other has been able to since the early twentieth century—combine general relativity and quantum theory into a single unified framework. This is an ambitious program that has occupied the best minds in mathematics and physics for decades. Einstein himself failed, but he lacked key ingredients that are necessary to pull it off.

String theory comes attached with a bit of controversy. As anyone who is reading this book likely knows, experimentally testing it is not an immediately accessible option due to the high energies required. It is, after all, a theory of creation itself—so the energies associated with string theory are of course very large. Nonetheless, it now appears that some indirect tests are possible and the timing of this book may coincide with some of this program. The first clue will be the continued search for *supersymmetry*, the theory that proposes fermions and bosons have *superpartners*, that is, a fermion like an electron has a sister superpartner particle that is a boson. Superparticles have not been discovered, so if it exists supersymmetry must be broken somehow so that the super partners have high mass. This could explain why we haven't seen them so far. But the Large Hadron Collider being constructed in Europe as we speak may be able to discover evidence of supersymmetry. This does not prove string theory, because you can have supersymmetry work just fine with point particles. However, supersymmetry is absolutely essential for string theory to work. If supersymmetry does not exist, string theory cannot be true. If supersymmetry is found, while it does not prove string theory, it is a good indication that string theory might be right.

Recent theoretical work also opens up the intriguing possibility that there might be large extra dimensions and that they might be inferred in experimental tests. Only gravity can travel into the extra space scientists call the "bulk." At the energies of the Large Hadron Collider, it might be possible to see some evidence that this is happening, and some have even proposed that microscopic black holes could be produced. Again, you could imagine having extra dimensions without string theory, so discoveries like these would not prove string theory. However, they would be major indirect evidence in its favor. You will learn in this book that string theory predicts the existence of extra dimensions, so any evidence of this has to be taken as a serious indication that string theory is on the right path.

String theory has lots of problems—it's a work in progress. This time is akin to living in the era when the existence of atoms was postulated but unproven and skeptics abounded. There are lots of skeptics out there. And string theory does seem a bit crazy—there are several versions of the theory, and each has a myriad of particle states that have not been discovered (however, note that transformations called *dualities* have been discovered that relate the different string theories, and work is underway on an underlying theory believed to exist called *M-theory*). The only serious competitor right now for string theory is loop quantum gravity. I want to emphasize I am not an expert, but I once took a seminar on it and to be honest I found it incredibly distasteful. It seemed so abstract it almost didn't seem like physics at all. It struck me more as mathematical philosophy. String theory seems a lot more physical to me. It makes outlandish predictions like the existence of extra dimensions, but general relativity and quantum theory make predictions that defy common sense as well. Eventually, all we can do is hope that experiment and observation will resolve the controversy and help us decide if loop quantum gravity or string theory is on the right track. Regardless of what our tastes are, since this is science we will have to follow where the evidence leads.

This book is written with the intent of getting readers started in string theory. It is intended for self-study and to make the real textbooks on the subject more accessible after you finish this one.

But make no mistake: This is not a "popular" book—it is written for readers who want to learn string theory.

The presentation has been simplified in some places. I have left out important topics like path integration, differential forms, and partition functions that are necessary for advanced study. Even so, there has been an attempt to give the reader a good overview of the basics of string physics. Unlike other introductory texts, I have decided to include a discussion of superstrings. It is more complicated, but my feeling is if you understand the bosonic case it's not too much of a leap to include superstrings. What you really need as background for this is some exposure to Dirac spinors. If you don't have this background, read Griffiths' *Elementary Particles* or try *Quantum Field Theory Demystified*. The bottom line is that string theory is an advanced topic, so you will need to have the background before reading this book. Specifically, from mathematics you need to know calculus, linear algebra, and partial and ordinary differential equations. It also helps to know some complex variables, and my book *Complex Variables Demystified* is being released at about the same time as this one to help you with this. This sounds like a long list and if you're just starting out it is. But you don't have to be an expert—just get a grasp of the topics and you should do fine with this book.

From physics, you should start off with wave motion if you're rusty with it. Open up a freshman physics book to do this. The core concepts you need for string theory are going to include wave motion on a string, boundary conditions on a string (from basic partial differential equations), the harmonic oscillator from quantum mechanics, and special relativity. Brush up on these before attempting to read this book. Due to limited space in the book, I did not include all of the background material from ground zero like Zweibach does in his fine text. I have attempted to present as accessible a presentation as possible but assume you already have done some background study. The three areas you need are quantum mechanics, relativity, and quantum field theory. Luckily there are three *Demystified* books available on these topics if you haven't studied them elsewhere.

In the short space allotted for a *Demystified* book, we can't cover everything from string theory. I have tried to strike a balance between building the basic physics and laying down the necessary mathematical machinery and being too advanced and introducing the most exciting topics. Unfortunately, this is not an easy program to pull off. I cover bosonic strings, superstrings, D-branes, black hole physics, and cosmology, among other topics. I have also included a discussion of the Randall-Sundrum model and how it resovles the hierarchy problem from particle physics.

I want to conclude by recommending Michio Kaku's popular physics books. I was actually "converted" from engineering to physics by reading one of his books that introduced me to the amazing world of string theory. It's hard to believe that picking up one of Kaku's books would have led me on a path such that I ended up writing a book on string theory. In any case, good luck on your quest to understand the universe, and I hope that this book makes that task more accessible to you.

David McMahon

CHAPTER 1

Introduction

General relativity and quantum mechanics stand out as the pillars of twentieth-century science, able to describe almost all known phenomena from the scale of subatomic particles all the way up to the rotations of galaxies and even the history of the universe itself. Despite this grand success, which includes stunning agreement with experiment, these two theories represent physics at a crossroads—one that is plagued with crisis and controversy.

The problem is that at first sight, these two theories are at complete odds with each other. The general theory of relativity (GR), Einstein's crowning achievement, describes gravitational interactions, that is, interactions that occur on the largest scales that we know. But it not only stands out as Einstein's greatest contribution to science but it also might be called the last classical theory of physics. That is, despite its revolutionary nature, GR does not take quantum mechanics into account at all. Since experiment indicates that quantum mechanics is the correct description for the behavior of matter, this is a serious flaw in the theory of general relativity.

We don't think about this under normal circumstances because quantum effects only become important in gravitational interactions that are extremely strong or taking place over very small scales. In the situations where we might apply general relativity, say to the motion of the planet mercury around the sun or the motion of the galaxies, quantum effects are not important at all. Two places where they will be important are in black hole physics and in the birth of the universe. We might also see quantum effects on gravity in very high energy particle interactions.

On the other hand, quantum mechanics basically ignores the insights of relativity. It basically pretends gravity doesn't exist at all, and pretends that space and time are not on the same footing. The notion of space-time does not enter in quantum mechanics, and although special relativity plays a central role in quantum field theory, gravitational interactions are nowhere to be found there either.

A Quick Overview of General Relativity

This isn't a book on GR, but we can give a very brief overview of the theory here (see *Relativity Demystified* for details). The central ideas of general relativity are the notion that geometry is dynamic and that the speed of light limits the speed of all interactions, including gravity. We start with the notion of the metric, which is a way of describing the distance between two points. In ordinary three-dimensional space the metric is

$$ds^2 = dx^2 + dy^2 + dz^2 \tag{1.1}$$

This metric follows from the pythagorean theorem by making the distances involved infinitesimal. Note that *this metric is invariant under rotations*. Something that is key to relativistic thinking is focusing on those quantities that are invariant.

To move up to a relativistic context, we extend the notion of a measure of distance between two points to a notion of distance between two *events* that happen in space *and* time. That is, we measure the distance between two points in *space-time*. This is done with the metric

$$ds^2 = -c^2 dt^2 + dx^2 + dy^2 + dz^2 \tag{1.2}$$

This metric extends the idea of geometry to include time as well. But not only that, it also extends the notion of a distance measure between two points that is invariant under rotations to one that is also invariant under Lorentz transformations, that is, Lorentz boosts between one inertial frame and another.

While adding time to the mix certainly extends the notion of geometry into an unfamiliar realm, we still have a fixed geometry that does not take into account gravitational fields. To extend the metric in a way that will do this, we have to enter the domain of *non-euclidean geometry*. This is geometry which does not require flat spaces. Instead we generalize to include spaces that are curved, like spheres or saddles. Now, since we are in a relativistic context, we need to include not just curved spaces but time as well, so we work with *curved space-time*. A general way to write Eq. (1.2) that will do this for us is

$$ds^2 = g_{\mu\nu}(x)dx^\mu dx^\nu \tag{1.3}$$

The *metric tensor* is the object $g_{\mu\nu}(x)$, which has components that depend on space-time. Now we have a dynamical geometry that varies from place to place and from time to time, and it turns out that $g_{\mu\nu}(x)$ is directly related to the gravitational field. Hence we arrive at the central truth of general relativity:

$$\text{gravity} \rightleftarrows \text{geometry}$$

Gravitational fields are essentially the geometry of space-time. The form of the metric tensor $g_{\mu\nu}(x)$ actually stems from the matter—energy that is present in a given region of space-time—which is the way that matter is the source of the gravitational field. The presence of matter alters the geometry, which changes the paths of free-falling particles giving the appearance of a gravitational field.

The equation that relates matter and geometry (i.e., the gravitational field) is called *Einstein's equation*. It has the form

$$R_{\mu\nu} + \frac{1}{2}g_{\mu\nu}R = 8\pi G T_{\mu\nu} \tag{1.4}$$

where $G = 6.673 \times 10^{-4}\, \text{m}^3/\text{kg}\cdot\text{s}^2$ is *Newton's constant of gravitation* and $T_{\mu\nu}$ is the *energy-momentum tensor*. $R_{\mu\nu}$ and R are objects that depend on the derivatives of the metric tensor $g_{\mu\nu}(x)$ and hence represent the dynamic nature of geometry in relativity. The energy-momentum tensor $T_{\mu\nu}$ tells us how much energy and matter is present in the space-time region being considered. The details of the equation are not important for our purposes, just keep in mind that matter (and energy) change the geometry of space-time giving rise to what we call a gravitational field, by changing the paths followed by free particles.

A Quick Primer on Quantum Theory

So matter enters the theory of relativity through the energy-momentum tensor $T_{\mu\nu}$. The rub is that we know that matter behaves according to the laws of quantum theory, which are at odds with the general theory of relativity. Without going into detail, we will review the basic ideas of quantum mechanics in this section (see *Quantum Mechanics Demystified* for a detailed description). In quantum mechanics, everything we could possibly find out about a particle is contained in the state of the particle or system described by a wave function:

$$\psi(x, t)$$

The wave function is a solution of Schrödinger's equation:

$$-\frac{\hbar^2}{2m}\nabla^2\psi + V\psi = i\hbar\frac{\partial\psi}{\partial t} \tag{1.5}$$

The wave function itself is not a real physical wave, rather it is a probability amplitude whose modulus squared $|\psi(x, t)|^2$ (note that the wave function can be complex) gives the probability that the particle or system is found in a given state.

Measurable observables like position and momentum are promoted to mathematical operators in quantum mechanics. They act on states (i.e., on wave functions) and must satisfy certain commutation rules. For example, position and momentum satisfy

$$[x, p] = i\hbar \tag{1.6}$$

Furthermore, there exists an *uncertainty principle* that puts constraints on the precision with which certain quantities can be known. Two important examples are

$$\Delta x\,\Delta p \geq \hbar/2$$
$$\Delta E\,\Delta t \geq \hbar/2 \tag{1.7}$$

So the more precisely we know the momentum of a particle, the less certain we are of its position and vice versa. The smaller the interval of time over which we examine a physical process, the greater the fluctuations in energy.

When considering a system with multiple particles, we have a wave function $\psi(x_1, x_2, \dots, x_n)$ say where there are n particles with coordinates x_i. It turns out that there are two basic types of particles depending on how the wave function behaves under particle interchange $x_i \rightleftarrows x_j$. Considering the two-particle case for simplicity, if the sign of the wave function is unchanged under

$$\psi(x_1, x_2) = \psi(x_2, x_1)$$

we say that the particles are *bosons*. Any number of bosons can exist in the same quantum state. On the other hand, if the exchange of two particles induces a minus sign in the wave function

$$\psi(x_1, x_2) = -\psi(x_2, x_1)$$

then we say that the particles in question are *fermions*. Fermions obey a constraint known as the Pauli exclusion principle, which says that no two fermions can occupy the same quantum state. So while boson number can assume any value $n_b = 0, \ldots, \infty$ the number of fermions that can occupy a quantum state is 0 or 1, that is, $n_f = 0, 1$ and not any other value.

The first move at bringing quantum theory and relativity together in the same framework is done by combining quantum mechanics together with the special theory of relativity (and hence leaving gravity out of the picture). The result, called *quantum field theory,* is a spectacular scientific success that agrees with all known experimental tests (see *Quantum Field Theory Demystified* for more details). In quantum field theory, space-time is filled with fields $\varphi(x, t)$ that act as operators. A given field can be Fourier expanded as

$$\varphi(x) = \int \frac{d^3k}{(2\pi)^{3/2}\sqrt{2\omega_k}} \left[\varphi(\vec{k})e^{-i(\omega_k x^0 - \vec{k}\cdot\vec{x})} + \varphi^*(\vec{k})e^{i(\omega_k x^0 - \vec{k}\cdot\vec{x})} \right]$$

We then express the fields in terms of creation and annihilation operators by making the transition $\varphi(\vec{k}) \to \hat{a}(\vec{k})$ and $\varphi^*(\vec{k}) \to \hat{a}^\dagger(\vec{k})$ giving

$$\hat{\varphi}(x) = \int \frac{d^3k}{(2\pi)^{3/2}\sqrt{2\omega_k}} \left[\hat{a}(\vec{k})e^{-i(\omega_k x^0 - \vec{k}\cdot\vec{x})} + \hat{a}^\dagger(\vec{k})e^{i(\omega_k x^0 - \vec{k}\cdot\vec{x})} \right]$$

The field then creates and destroys particles that are the *quanta* of the given field.

We require that all quantities be Lorentz invariant. To get quantum theory more into the picture, we impose commutation relations on the fields and their conjugate momenta

$$\left[\hat{\varphi}(x), \hat{\pi}(y) \right] = i\delta(\vec{x} - \vec{y})$$

$$\left[\hat{\varphi}(x), \hat{\varphi}(y) \right] = 0$$

$$\left[\hat{\pi}(x), \hat{\pi}(y) \right] = 0$$

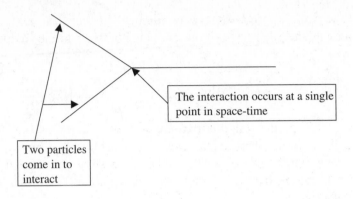

Figure 1.1   In particle physics, interactions occur at a single point.

where $\hat{\pi}(x)$ is the conjugate momenta obtained from the field $\varphi(x,\ t)$ using standard techniques from lagrangian mechanics.

From the commutation relations and from the form of the fields (i.e., the creation and annihilation operators) you might glean that particle interactions take place *at specific, individual points in space-time*. This is important because it means that particle interactions take place over zero distance. Particles in quantum field theory are point particles represented mathematically as located at a single point. This is illustrated schematically in Fig. 1.1.

Now, calculations in quantum field theory can be done using a perturbative expansion. Each term in the expansion describes a possible particle interaction and it can be represented graphically using a *Feynman diagram*. For example, in Fig. 1.2, we see two electrons scattering off each other.

The Feynman diagram in Fig. 1.2 represents the lowest-order term in the series describing the amplitude for the process to occur. Taking more terms in the series, we add diagrams with more complex internal interactions that have the same

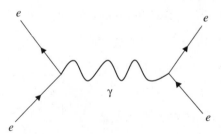

Figure 1.2   A Feynman diagram illustrating the scattering of two electrons.

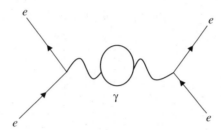

Figure 1.3   The photon turns into an electron-positron pair (the circle) that
subsequently annihilate producing another photon.

initial and final states. For example, the exchanged photon might turn into an
electron-positron pair, which subsequently decays into another photon. This is
illustrated in Fig. 1.3.

Interior processes like that shown in Fig. 1.3 are called *virtual*. This is because
they do not appear as initial or final states. To find the actual amplitude for a given
process to occur, we need to draw Feynman diagrams for every possible virtual
process, that is, take all the terms in the series. In practice we can take only as many
terms as we need to get the accuracy desired in our calculations.

This type of procedure works well in the electromagnetic, weak, and strong
interactions. However, the overall procedure has some big problems and they
cannot be dealt with when gravity is involved. The problem comes down to the fact
that interactions occur at a single space-time point. This leads to infinite results in
calculations (aptly called *infinities*). Technically speaking, the calculation of a
given amplitude which includes all virtual processes involves an integral over all
possible values of momentum. This can be described by a *loop integral* that can be
written in the form

$$I \sim \int p^{4J-8} d^D p \tag{1.8}$$

Here p is momentum, J is the spin of the particle, and D, which is seen in the
integration measure, is the dimension of space-time. Now consider the quantity

$$\lambda = 4J + D - 8 \tag{1.9}$$

If the momentum $p \to \infty$ and

$$\lambda < 0$$

then I in Eq. (1.8) is finite and calculations give answers that make sense. On the other hand, if the momentum $p \to \infty$ but

$$\lambda > 0$$

the integral in Eq. (1.8) diverges. This leads to infinities in calculations. Now if $I \to \infty$ but does so slowly, then a mathematical technique called *renormalization* can be used to get finite results from calculations. Such is the case when working with established theories like quantum electrodynamics.

The Standard Model

In its finished form, the theoretical framework that describes known particle interactions with quantum field theory is called the *standard model*. In the standard model, there are three basic types of particle interactions. These are

- Electromagnetic
- Weak
- Strong (nuclear)

There are two basic types of particles in the standard model. These are

- Spin-1 gauge bosons that transmit particle interactions (they "carry" the force). These include the photon (electromagnetic interactions), $W^{\pm}$ and Z (weak interactions), and gluons (strong interactions).
- Matter is made out of spin-1/2 fermions, such as electrons.

In addition, the standard model requires the introduction of a spin-0 particle called the *Higgs boson*. Particles interact with the associated Higgs field, and this interaction gives particles their mass.

Quantizing the Gravitational Field

The general theory of relativity includes gravitational waves. They carry angular momentum $J = 2$, so we deduce that the quantum of the gravitational field, known as the *graviton*, is a spin-2 particle. It turns out that string theory naturally includes a spin-2 boson, and so naturally includes the quantum of gravity. Returning to Eq. (1.8), if we let $J = 2$ and consider space-time as we know it $D = 4$, then

$$4J - 8 + D = 4(2) - 8 + 4 = 4$$

So in the case of the graviton,

$$p^{4J-8} \to p^0 = 1$$

and

$$I \sim \int d^4 p \to \infty$$

when integrated over all momenta. This means that gravity cannot be renormalized in the way that a theory like quantum electrodynamics can, because it diverges like p^4. In contrast, consider quantum electrodynamics. The spin of the photon is 1, so

$$4J - 8 + D = 4(1) - 8 + 4 = 0$$

and the loop integral goes as

$$I \sim \int p^{4J-8} d^D p = \int p^{-4} d^4 p \Rightarrow$$

Goes like

$$\approx p^0 = 1$$

This tells us that incorporating gravity into the standard quantum field theory framework is an extremely problematic enterprise. The bottom line is nobody really had any idea how to do it for a very long time.

String theory gets rid of this problem by getting rid of particle interactions that occur at a single point. Take a look at the uncertainty principle

$$\Delta x \, \Delta p \sim \hbar$$

If momentum blow up, that is, $\Delta p \to \infty$, this implies that $\Delta x \to 0$. That is, large (infinite) momentum means small (zero) distance. Or put another way, pointlike

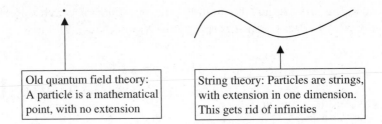

> Old quantum field theory:
> A particle is a mathematical
> point, with no extension

> String theory: Particles are strings,
> with extension in one dimension.
> This gets rid of infinities

Figure 1.4 In string theory, particles are replaced by strings, spreading out interactions over space-time so that infinities don't result.

interactions (zero distance) imply infinite momentum. This leads to divergent loop integrals, and infinities in calculations.

So in string theory, we replace a point particle by a one-dimensional string. This is illustrated in Fig. 1.4.

Some Basic Analysis in String Theory

In string theory, we don't go all the way to $\Delta x \to 0$ but instead cut it off at some small, but nonzero value. This means that there will be an upper limit to momentum and hence $\Delta p \not\to \infty$. Instead momentum goes to a large, but finite value and the loop integral divergences can be gotten rid of.

If we have a cutoff defined by the length of a string, then the uncertainty relations must be modified. It is found that in a string theory uncertainty in position Δx is approximately given by

$$\Delta x = \frac{\hbar}{\Delta p} + \alpha' \frac{\Delta p}{\hbar} \qquad (1.10)$$

A new term has been introduced into the uncertainty relation, $\alpha'(\Delta p/\hbar)$ which can serve to fix a minimum distance that exists in the theory. The parameter α' is related to the *string tension* T_S as

$$\alpha' = \frac{1}{2\pi T_S} \qquad (1.11)$$

The minimum distance scale we can see in string theory is given by

$$x_{\min} \sim 2\sqrt{\alpha'} \qquad (1.12)$$

So if $\alpha' \neq 0$, then the problems that result from pointlike interactions are avoided because they cannot take place. Interactions are spread out and infinities are avoided.

Unification and Fundamental Constants

String theory proposes to be a unified theory of physics. That is, it is supposed to be the most fundamental theory that describes all particle interactions (known and perhaps currently unknown), particle types, and gravity. We can gain some insight into the unification of all forces into a single framework by building up quantities from the fundamental constants in the theory.

If you have studied quantum field theory then you know that a dimensionless constant called the fine structure constant can be constructed out of e, $\hbar$, and c where e is the charge of the electron, $\hbar$ is Planck's constant, and c is the speed of light. The fine structure constant α gives us a measure of the strength of the electromagnetic field (through the coupling constant). It is given by

$$\alpha_{EM} = \frac{e^2}{4\pi\hbar c} \approx \frac{1}{137} < 1 \qquad (1.13)$$

The fact that $\alpha_{EM} < 1$ is what makes perturbation theory possible, since we can expand a quantity in powers of α_{EM} to obtain approximate answers.

A similar procedure can be applied to gravity. We consider the gravitational force because it is the only force not described in a unified framework based on quantum theory. The other known forces, the electromagnetic, the weak, and the strong forces are described by the standard model, while gravity sits on the sidelines relegated to the second string classical team. The constants important in gravitational interactions include Newton's gravitational constant G, the speed of light c, and if we are talking about a quantum theory of gravity, then we need to include Planck's constant $\hbar$. Two fundamental quantities can be derived using these constants, a *length* and a *mass*. This tells us the distance and energy scales over which quantum gravity will start to become important.

First let's consider the length, which is aptly called the *Planck length*. It is given by

$$l_p = \sqrt{\frac{G\hbar}{c^3}} \sim 10^{-35} \text{ m} \qquad (1.14)$$

This is one very small distance. For comparison, the dimension of a typical atomic nucleus is

$$l_{nucleus} \sim 10^{-15} \text{ m}$$

which is bigger than the Planck length by a factor of 10^{20}! This means that quantum gravitational effects can (naively at least) be expected to take place over very small distance scales. To probe such small distance scales, you need very high energies. This is confirmed by computing the Planck mass, which turns out to be given by

$$M_p = \sqrt{\frac{\hbar}{Gc}} \sim 10^{-8} \text{ kg} \qquad (1.15)$$

While this is a small value to what might be measured when considering your waistline, it's pretty large compared to the masses of the fundamental particles. This tells us, again, that high energies are needed to probe the realm of quantum gravity. The Planck mass also turns out to be the mass of a black hole where its Schwarzschild radius is the same as its Compton wavelength, suggesting that this is a length scale at which quantum gravitational effects become significant.

Next we can form a *Planck time*. This is given by

$$t_p = \frac{l_p}{c} \sim 10^{-44} \text{ s} \qquad (1.16)$$

This is a small time interval indeed. So if you think quantum gravity, think small distances, small time intervals, and large energies. At these high energies gravity becomes strong. To see how this works think about the following. In a freshman physics course you learn that the electromagnetic force is something like 10^{40} times as strong as the gravitational interaction. But at the high energies we are describing, where quantum gravity becomes important, the strength of gravitational interactions is comparable to that of the other forces—gravity becomes strong and hence is important in particle interactions. Since the particle accelerators that are currently in existence (or that can even be dreamed up) probe energies that fall on a much smaller scale, gravity can be considered to be extremely weak at presently accessible energies.

String Theory Overview

So far we've seen why strings can be useful in developing a finite quantum theory of gravity, and we've seen the energy scales over which such a theory might be important. Let's close the chapter by looking at some basic notions included in string theory. The first is that fundamental particles are not points, they are strings, as shown in Fig. 1.5.

Figure 1.5 Fundamental particles are extended one-dimensional objects called strings.

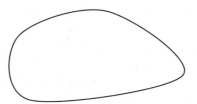

Figure 1.6 A closed string has no loose ends.

Strings can be *open* (Fig. 1.5) or *closed* (Fig. 1.6), the latter meaning that the ends are connected.

Excitations of the string give different fundamental particles. As a particle moves through space-time, it traces out a world line. As a string moves through space-time, it traces out a worldsheet (see Fig. 1.7), which is a surface in space-time parameterized by (σ, τ). A mapping $x^\mu(\tau, \sigma)$ maps a worldsheet coordinate (σ, τ) to the space-time coordinate x.

So, in the world according to string theory, the fundamental objects are tiny strings with a length on the order of the Planck scale (10^{-33} cm). Like any string,

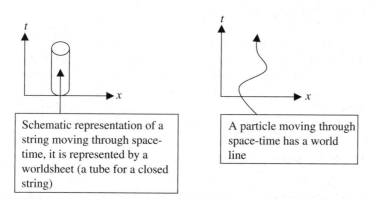

Figure 1.7 A comparison of a worldsheet for a closed string and a world line for a point particle. The space-time coordinates of the world line are parameterized as $x^\mu = x^\mu(\tau)$, while the space-time coordinates of the worldsheet are parameterized as $x^\mu(\tau, \sigma)$ where (σ, τ) give the coordinates on the surface of the worldsheet.

these fundamental strings can vibrate and vibrations at different resonant frequencies (excitations of the string) give rise to particles with different properties. For a particle with spin J and mass m_J, the mass and spin of the particle are related to the string tension through α' as

$$J = \alpha' m_J^2 \tag{1.17}$$

Think of a vibrating string having different modes in the way that a violin string can vibrate at different frequencies. Instead of having a plethora of "fundamental particles" with mysterious origin, there is only one fundamental object—a string that vibrates with different modes giving the appearance that there are multiple fundamental objects. Each mode appears as a different particle, so one mode could be an electron, while another, different mode could be a quark.

It is possible for strings to split apart and to combine. Let's focus on strings splitting apart. Suppose that a parent string is vibrating in a mode corresponding to particle A. It splits in two, with resulting daughter strings vibrating in modes corresponding to particles B and C respectively. This process of splitting corresponds to the particle decay:

$$A \rightarrow B + C$$

Conversely, strings can join up as well, combining to form a single string. This is a process that until now we have thought of as particle absorption. So processes that seemed more on the mysterious side, such as particle decay, are explained with a simple conceptual framework.

TYPES OF STRING THEORIES

There appear to be five different types of string theory, but it has been shown that they are different ways of looking at the same theory, with the different types related by *dualities*. The five basic types are

- **Bosonic string theory**   This is a formulation of string theory that only has bosons. There is no supersymmetry, and since there are no fermions in the theory it cannot describe matter. So it is really just a toy theory. It includes both open and closed strings and it requires 26 space-time dimensions for consistency.

- **Type I string theory**   This version of string theory includes both bosons and fermions. Particle interactions include supersymmetry and a gauge group $SO(32)$. This theory and all that follow require 10 space-time dimensions for consistency.

- **Type II-A string theory**   This version of string theory also includes supersymmetry, and open and closed strings. Open strings in type II-A string theory have their ends attached to higher-dimensional objects called *D-Branes*. Fermions in this theory are not chiral.

- **Type II-B string theory**   Like type II-A string theory, but it has chiral fermions.

- **Heterotic string theory**   Includes supersymmetry and only allows closed strings. Has a gauge group called $E_8 \times E_8$. The left- and right-moving modes on the string actually require different numbers of space-time dimensions (10 and 26). We will see later that there are actually two heterotic string theories.

M-THEORY

All these string theories might seem confusing, and make the whole enterprise seem like a stab in the dark. However, as we go through the book we will learn about the different dualities that connect the different types of string theories. These go by the names of *S duality* and *T duality*.

Since these dualities exist, there has been speculation that there is an underlying, more fundamental theory. It does by the odd name of *M-theory* but "M" does not really have any agreed upon or specific meaning (perhaps mother of all theories). One concept in M-theory is that the space-time manifold (i.e., its structure) is not assumed a priori but rather emerges from the vacuum.

One concrete manifestation of M-theory is based on matrix mechanics, the kind you are used to from ordinary quantum mechanics. In this context "M" really means something, and we call it *matrix theory*. In this theory, if we *compactify* (i.e., make really tiny) n spatial dimensions on a torus, we get out a dual matrix theory that is just an ordinary quantum field theory in $n + 1$ space-time dimensions.

D-BRANES

A *D-brane*, mentioned in our discussion of string theory types, is an extension of the common sense notion of a membrane, which is a two-dimensional brane or 2-brane. A string can be though of as a one-dimensional brane or 1-brane. So a *p-brane* is an object with p spatial dimensions.

D-branes are important in string theory because the ends of fundamental strings can attach to them. It is believed that quantum fields described by Yang-Mills type theories (such as electromagnetism) involve strings that are attached by D-branes. This idea has great explanatory power, because gravitons, the quantum of gravity, are *not* attached to D-branes. They can travel or "leak off" a D-brane, so we don't see as many of them. This explains what until now has been a great mystery, why electromagnetism (and the other known forces) is so much stronger than gravity.

So this picture of the universe has a three-dimensional brane (or 3D-brane) embedded in a higher-dimensional space-time called the *bulk*. Since we interact with the physical world primarily through electromagnetic forces (light, chemical reactions, etc.), which are mediated by particles that are really strings stuck to the brane, we experience the world as having three spatial dimensions. Gravity is mediated by strings that can leave the brane and travel off into the bulk, so we see it as a much weaker force. If we could probe the bulk somehow, we would see that gravity is actually comparable in strength.

HIGHER DIMENSIONS

We live in a world with three spatial dimensions. In a nutshell this means that there are three distinct directions through which movement is possible: up-down, left-right, and forward-backward. In addition, we have the flow of time (forward only as far as we know). Mathematically, this gives us the relativistic description of coordinates (x, y, z, t).

It is possible to imagine a world where one of the spatial directions or dimensions have been removed (say up-down). Such a two-dimensional world was described by Edward Abbott in his classic *Flatland*. What if instead, we added dimensions? This idea is actually pretty useful in physics, because it provides a pathway toward unifying different physical theories. This kind of thinking was originally put forward by two physicists named Kaluza and Klein in the 1920s. Their idea was to bring gravity and electromagnetism into a single theoretical framework by imagining that these two theories were four-dimensional limits of a five-dimensional supertheory. This idea did not work out, because back then people did not know about quantum field theory and so did not have a complete picture of particle interactions, and did not know that the fully correct description of electromagnetic interactions is provided by quantum electrodynamics. But this idea has a lot of appeal and reemerged in string theory.

Kaluza and Klein had to explain why we don't see the higher dimension, and hit upon the idea of compactification—a procedure where we make the higher dimensions so small they are not detectable at lower energy (i.e., on the kind of energy scales that we live in). If they are small enough, the extra dimensions can't be noticed or detected scientifically without the existence of the appropriate technology. If they are so small that they are on the Planck scale, we might not be able to see them at all. This concept is illustrated in Fig. 1.8.

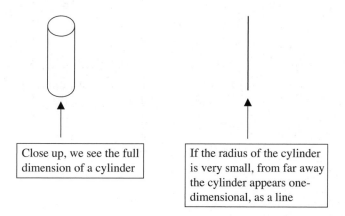

Figure 1.8   Compactification explains why we may not be aware of extra spatial dimensions even if they exist. If the radius of a cylinder is very small, from far away it looks like a line.

String theory requires the existence of extra spatial dimensions for technical reasons that we will discuss in later chapters. An interesting side effect of these extra dimensions is that another mystery of particle physics is done away with. Experimentalists have worked out that there are three *families* of particles. For example, when considering leptons, there is the electron and its corresponding neutrino. But there are also the "heavy electrons" known as the muon and the tau, together with their corresponding neutrinos, that are really just duplicates of the electron. The same situation exists for the quarks. Why are there three particle families? And why are there the types of particle interactions that we see? It turns out that higher spatial dimensions together with string theory may provide an answer.

The way that you compactify the extra dimensions (the *topology*) determines the numbers and types of particles seen in the universe. In string theory this results from the way that the strings can wrap around the compactified dimensions, determining what vibrational modes are possible in the string and hence what types of particles are possible.

One important compactified manifold that we will see is called the *Calabi-Yau* manifold. A Calabi-Yau manifold that compactifies six spatial dimensions and leaves three spatial dimensions "macroscopic" plus time gives a ten-dimensional universe as required by most of the string theories. A key aspect of Calabi-Yau manifolds is that they *break* symmetries. Thus another mystery of particle physics is explained, so-called spontaneous symmetry breaking (see *Quantum Field Theory Demystified* for a description of symmetry breaking).

Summary

Quantum mechanics and general relativity were the major developments in theoretical physics in the twentieth century. Unifying them into a single theoretical framework has proven extremely challenging, if not impossible. This is because the resulting quantum theories are plagued by infinities that result from the fact that interactions take place at a single mathematical point (zero distance scale). By spreading out the interactions, string theory offers the hope of developing not only a unified theory of particle physics, but a finite theory of quantum gravity.

Quiz

1. If $\lambda = 4J + D - 8 > 0$ and $p \to \infty$ then
    - (a) the loop integral is convergent.
    - (b) the loop integral diverges.
    - (c) the loop integral can be calculated, but the results are meaningless.

2. The scale of the Planck length and Planck mass tell us that quantum gravity
    - (a) operates on small-distance and high-energy scales.
    - (b) is nonsensical.
    - (c) operates on small-distance and small-energy scales.
    - (d) operates on large-distance and small-energy scales.

3. Perturbation theory is possible in quantum electrodynamics because
    - (a) $\alpha_{EM} > 1$
    - (b) $\alpha_{EM} = 1$
    - (c) $\alpha_{EM} < 1$
    - (d) Perturbation theory is not possible in quantum electrodynamics

4. The quantum uncertainty relations are modified in string theory as
    - (a) $\Delta x \sim \dfrac{\hbar}{\Delta p} + \dfrac{\Delta p}{\hbar}$
    - (b) $\Delta x \sim \dfrac{\hbar}{\Delta p} + \alpha' \dfrac{\Delta p}{\hbar}$
    - (c) $\Delta x \sim \dfrac{\hbar}{\Delta p} - \alpha' \dfrac{\Delta p}{\hbar}$
    - (d) $\Delta x \sim \dfrac{\hbar}{\alpha' \Delta p}$

5. The minimum distance scale in string theory is about

   (a) $x_{min} \sim \dfrac{1}{2\sqrt{\alpha'}}$

   (b) $x_{min} \sim 2\sqrt{T_S}$

   (c) $x_{min} \sim 0$

   (d) $x_{min} = \sqrt{\dfrac{2}{\pi T_S}}$

6. The topology of compactified dimensions

   (a) determines the types of particles seen in the universe.

   (b) has no impact on particle interactions.

   (c) restores symmetries in quantum field theories.

7. Heterotic string theory has the gauge group

   (a) $E_6 \times E_6$

   (b) $SU(32)$

   (c) $E_8 \times E_8$

   (d) $SO(16)$

8. String theory explains the difference between electromagnetism and gravity as

   (a) String theory provides the unification energy of gravity and electromagnetism.

   (b) Gravitons are not connected to the brane, so can leak off into the bulk making gravity appear much weaker than electromagnetism.

   (c) Photons leak off into the bulk, making electromagnetic phenomena more prominent.

9. Bosonic string theory is not realisitic because

   (a) it includes 26 space-time dimensions.

   (b) it does not allow Calabi-Yau compactification.

   (c) it does not include fermions, so cannot describe matter.

   (d) it lacks a $E_8 \times E_8$ symmetry group.

10. In string theory particle decay is explained by

   (a) a string splitting apart into multiple daughter strings.

   (b) it remains poorly understood.

   (c) quantum tunneling through the string potential.

   (d) strong vibrational modes that decouple the string.

CHAPTER 2

The Classical String I: Equations of Motion

When you studied classical mechanics and quantum field theory, you learned about the action and deriving the equations of motion from the Euler-Lagrange equations. This can be done in the case of the string, and it can be done relativistically. If we are going to consider a unified theory of physics, this is a good place to start—ensuring that we understand how to describe the dynamics of strings in a manner that is fully consistent with relativity before moving on to introduce the quantum theory.

When we quantize our strings, our first foray into a fully relativistic, quantum theory will be an instructive but unrealistic case, the *bosonic string*. As the name implies, we are going to look at a theory consisting exclusively of bosons—that is, states with *integral* spin. We know that this cannot be a realistic theory because in the actual universe while force-carrying particles are indeed bosons, fundamental matter particles (like electrons) have *half-integral spin*, that is, they are *fermions*. So a theory that describes a world consisting entirely of bosons does not describe the real universe.

Nonetheless, we start here because it is an easier way to approach string theory and we can learn the nuts and bolts in a slightly simpler context. We are going to approach bosonic strings in three steps. In this chapter, we will develop the theory of classical, relativistic strings starting with the action principle and deriving the equations of motion. In Chap. 3, we will learn about the stress-energy tensor and conserved currents, specifically *conserved worldsheet currents*. Finally, in the last chapter of this part of the book, we will quantize the strings using a procedure of first quantization (i.e., first quantization of point particles gives single-particle states). In the end you have a quantized relativistic theory.

To this end, we begin our journey into the world of classical relativistic point particle moving in space-time to illustrate the techniques used.

The Relativistic Point Particle

The task at hand is to describe the motion of a free (relativistic) point particle in space-time. One way to approach the problem is by using an *action principle*. Before we do that, let's set up the arena in which the particle moves. Let its motion be defined with respect to space-time coordinates X^μ where X^0 is the timelike coordinate (i.e., $X^0 = ct$) and X^i where $i \neq 0$ are the spacelike coordinates (say x, y, and z). While you are probably used to lowercase letters like x^μ to represent coordinates, in string theory uppercase letters are used, so we will stick to that convention.

Anticipating the fact that string theory takes place in a higher-dimensional arena, rather than the usual one time dimension and three spatial dimensions we are used to, we consider motion in a D-dimensional space-time. There is one time dimension but now we allow for the possibility of $d = D - 1$ spatial dimensions. We reserve 0 to index the time dimension hence our coordinates range over $\mu = 0, \ldots, d$.

Now, the motion or trajectory of a particle is described such that the coordinates are parameterized by τ, which parameterizes the world-line of the particle. That is, this is the time given by a clock that is moving or carried along with the particle itself. We can emphasize this parameterization by writing the coordinates as functions of the proper time:

$$X^\mu = X^\mu(\tau) \qquad (2.1)$$

To describe distance measurements, we are going to need a *metric*, that is, a function which allows us to define the distance between two points. Here we will stick with special relativity and use the flat space Minkowski metric which is usually denoted by $\eta_{\mu\nu}$. You may recall that the time and spatial components of the metric have different sign; the choice used is referred to as the *signature* of the metric. In string theory, it is convenient to place the negative sign with the time component, so in the case of $d = 3$ spatial dimensions we can write the Minkowski metric as a matrix

$$\eta_{\mu\nu} = \begin{pmatrix} -1 & 0 & 0 & 0 \\ 0 & 1 & 0 & 0 \\ 0 & 0 & 1 & 0 \\ 0 & 0 & 0 & 1 \end{pmatrix} \tag{2.2}$$

More compactly, we can write $\eta_{\mu\nu} = (-, +, +, +)$. Generalizing to D-dimensional Minkowski space-time, we simply associate a plus sign with products of spatial coordinates. So the Lorentz invariant length squared of a vector is

$$(X^{\mu})^2 = \eta_{\mu\nu} X^{\mu} X^{\nu} = -(X^0)^2 + (X^1)^2 + \cdots + (X^d)^2 \tag{2.3}$$

Infinitesimal lengths or distances are described by

$$ds^2 = -\eta_{\mu\nu} dX^{\mu} dX^{\nu} = (dX^0)^2 - (dX^1)^2 - \cdots (dX^d)^2 \tag{2.4}$$

We include the minus sign out in front of the metric in Eq. (2.4) to ensure that $ds = \sqrt{-\eta_{\mu\nu} dX^{\mu} dX^{\nu}}$ is real for timelike trajectories. With these notations in hand, we are ready to describe the trajectory of a free relativistic particle using the action principle.

The action principle tells us that the relativistic motion of a free particle is proportional to the invariant length of the particles trajectory. That is,

$$S = -\alpha \int ds \tag{2.5}$$

First let's figure out what the constant of proportionality is.

EXAMPLE 2.1
Given that the action of a free, non-relativistic particle is $S_0 = \int dt\,(1/2)mv^2$, where m is the mass of the particle and v is the particle velocity, determine the nature of the constant in Eq. (2.5).

SOLUTION

For simplicity, we consider motion in one spatial dimension. Now

$$S = -\alpha \int ds$$

$$= -\alpha \int \sqrt{dt^2 - dx^2}$$

$$= -\alpha \int dt \sqrt{1 - \frac{dx^2}{dt^2}}$$

$$= -\alpha \int dt \sqrt{1 - v^2}$$

Now recall the binomial theorem. This tells us that

$$\sqrt{1 \pm x} \approx 1 \pm \frac{1}{2} x$$

Hence,

$$\sqrt{1 - v^2} \approx 1 - \frac{1}{2} v^2$$

Therefore

$$S = -\alpha \int dt \sqrt{1 - v^2}$$

$$\approx -\alpha \int dt \left(1 - \frac{1}{2} v^2\right) = -\alpha \int dt + \int dt \frac{1}{2} \alpha v^2$$

Comparison of the second term in this expression with $S_0 = \int dt \, (1/2) m v^2$ tells us that α must be the mass of the particle.

We can also determine the units of α and deduce that it is the mass of the particle from dimensional analysis. First, what are the units of action? Recall from your studies of quantum theory that the units of action from Planck's constant $\hbar$ are mass times length squared per time:

$$[\hbar] = \frac{ML^2}{T} \tag{2.6}$$

Now let's look at $S = -\alpha \int ds$. From the integral, we have length L, so we have

$$\frac{ML^2}{T} = [\alpha]L$$

So it must be the case that

$$[\alpha] = \frac{ML}{T}$$

We can obtain this result using the mass of the particle together with the speed of light c, which is of course a length over time. That is,

$$\alpha = \frac{m}{c}$$

$$\Rightarrow [\alpha] = \frac{ML}{T}$$

In units where $c = \hbar = 1$, which are commonly used in particle physics and string theory, the action is dimensionless. Hence mass is inverse length and

$$\alpha = m$$

$$\Rightarrow [\alpha] = M = \frac{1}{L} \tag{2.7}$$

Now let's see how to write down the action and obtain the equations of motion from it. We start with the definition of infinitesimal length given in Eq. (2.4). This gives the action as

$$S = -m \int \sqrt{-\eta_{\mu\nu} dX^\mu dX^\nu} \tag{2.8}$$

Let's rewrite the integrand:

$$\sqrt{-\eta_{\mu\nu} dX^\mu dX^\nu} = \sqrt{-\eta_{\mu\nu} \left(\frac{d\tau}{d\tau}\right)^2 dX^\mu dX^\nu}$$

$$= d\tau \sqrt{-\eta_{\mu\nu} \frac{dX^\mu}{d\tau} \frac{dX^\nu}{d\tau}} = d\tau \sqrt{-\eta_{\mu\nu} \dot{X}^\mu \dot{X}^\nu}$$

This allows us to write the action in the form

$$S = -m \int d\tau \sqrt{-\eta_{\mu\nu} \dot{X}^{\mu} \dot{X}^{\nu}} \tag{2.9}$$

This action is a nice compact form that allows us to derive the equations of motion. As you recall from your studies of classical mechanics or quantum field theory, the quantity in the integrand is called the *lagrangian*:

$$L = -m \sqrt{-\eta_{\mu\nu} \dot{X}^{\mu} \dot{X}^{\nu}} = -m \sqrt{-\dot{X}^2}$$

There are two problems with the action so far developed in Eq. (2.9). First, think about what happens in the case of a massless particle. Setting $m = 0$ leaves us with $S \to 0$ and so there is nothing to vary to obtain the equations of motion. So this action isn't very helpful in the case of a massless particle. Also, it turns out that quantization is not easy when we have a square root in the action. For these reasons, we introduce an *auxiliary field* that we will denote $a(\tau)$ and consider the lagrangian

$$L = \frac{1}{2a} \dot{X}^2 - \frac{m^2}{2} a$$

We can use this to define an alternative expression for the action

$$S' = \frac{1}{2} \int d\tau \left(\frac{1}{a} \dot{X}^2 - m^2 a \right) \tag{2.10}$$

We can vary this action to find an equation of motion for the auxiliary field $a(\tau)$. We find

$$\delta S' = \frac{1}{2} \delta \int d\tau \left(\frac{1}{a} \dot{X}^2 - m^2 a \right)$$

$$= \frac{1}{2} \int d\tau \left(\delta \left(\frac{1}{a} \right) \dot{X}^2 - \delta(m^2 a) \right)$$

$$= \frac{1}{2} \int d\tau \left(-\frac{1}{a^2} \dot{X}^2 - m^2 \right)$$

Now we set $\delta S' = 0$. Since this means that the integrand must be 0, we obtain the equation

$$-\frac{1}{a^2}\dot{X}^2 - m^2 = 0$$

$$\Rightarrow \dot{X}^2 + m^2 a^2 = 0$$

This is the equation of motion for the auxiliary field. From this we find an expression for the auxiliary field given by

$$a = \sqrt{-\frac{\dot{X}^2}{m^2}} \qquad\qquad (2.11)$$

Using Eq. (2.11), we can show that the action in Eq. (2.10) is equivalent to Eq. (2.8), which we do in the next example.

EXAMPLE 2.2
Show that if $a = (-\dot{X}^2/m^2)^{1/2}$, the action $S' = 1/2\int d\tau[(1/a)\dot{X}^2 - m^2 a]$ can be recast in the form $S = -m\int \sqrt{-\eta_{\mu\nu}dX^\mu dX^\nu}$.

SOLUTION
Let's start by recalling that $\dot{X}^2 = \eta_{\mu\nu}\dot{X}^\mu \dot{X}^\nu$. So we can rewrite the action S' in the following way:

$$S' = \frac{1}{2}\int d\tau\left(\frac{1}{a}\dot{X}^2 - m^2 a\right)$$

$$= \frac{1}{2}\int d\tau\left(\sqrt{\frac{-m^2}{\dot{X}^2}}\dot{X}^2 - m^2\sqrt{-\frac{\dot{X}^2}{m^2}}\right)$$

$$= \frac{1}{2}\int d\tau\left(\sqrt{\frac{-m^2}{\dot{X}^2}}\dot{X}^2 - m\sqrt{-\dot{X}^2}\right)$$

$$= \frac{1}{2}\int d\tau\left(\sqrt{\frac{-m^2}{\dot{X}^2}}\dot{X}^2 - m\sqrt{-\eta_{\mu\nu}\dot{X}^\mu \dot{X}^\nu}\right)$$

Now, let's use a simple algebraic trick to rewrite the first term. Remember from complex variables that $i^2 = -1$. This means that

$$\sqrt{\frac{-m^2}{\dot{X}^2}\dot{X}^2} = (-1)(-1)\sqrt{\frac{-m^2}{\dot{X}^2}\dot{X}^2}$$

$$= -\sqrt{\frac{-m^2}{\dot{X}^2}i^2\dot{X}^2}$$

$$= -\sqrt{\frac{-m^2}{\dot{X}^2}i^4\dot{X}^4}$$

$$= -\sqrt{-m^2 i^4 \dot{X}^2}$$

$$= -m\sqrt{-i^4\dot{X}^2}$$

But $i^4 = +1$, and so $-m\sqrt{-i^4\dot{X}^2} = -m\sqrt{-\dot{X}^2} = -m\sqrt{-\eta_{\mu\nu}\dot{X}^\mu\dot{X}^\nu}$. Therefore the action is

$$S' = \frac{1}{2}\int d\tau\left(\sqrt{\frac{-m^2}{\dot{X}^2}\dot{X}^2} - m\sqrt{-\eta_{\mu\nu}\dot{X}^\mu\dot{X}^\nu}\right)$$

$$= \frac{1}{2}\int d\tau\left(m\sqrt{-\eta_{\mu\nu}\dot{X}^\mu\dot{X}^\nu} - m\sqrt{-\eta_{\mu\nu}\dot{X}^\mu\dot{X}^\nu}\right)$$

$$= -m\int d\tau\left(\sqrt{-\eta_{\mu\nu}\dot{X}^\mu\dot{X}^\nu}\right) = S$$

This demonstrates that the two actions are equivalent.

Strings in Space-Time

At this point we have reviewed some basic techniques that help us calculate the equations of motion for a free relativistic point particle. We are going to extend this work to the case of a string moving in space-time. A point particle has no extent whatsoever, so can be described as a zero-dimensional object. We have seen that its

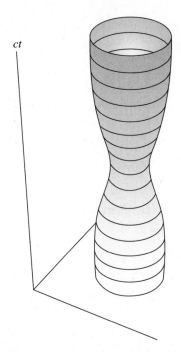

Figure 2.1   The world-sheet of a closed string is a tube in space-time.

motion can be described by saying that a point particle (zero-dimensional) sweeps out a path or line in space-time (one dimension) that we call the world-line. A string, unlike a point particle, has some extension in one dimension, so it's a one-dimensional object. As it moves, the string (one-dimensional) sweeps out a two-dimensional surface in space-time that scientists call a *worldsheet*. For example, imagine a closed loop of string moving through space-time. The worldsheet in this case will be a *tube*, as shown in Fig. 2.1.

We can summarize this in the following way:

- The path of a point particle is a line in space-time. A line can be parameterized by a single parameter, which is the proper time.

- As a string moves through space-time it sweeps out a two-dimensional surface called a worldsheet. Since the worldsheet is two-dimensional, we need two parameters, which we can generally denote by ξ_1 and ξ_2.

Locally the coordinates ξ_1 and ξ_2 can be thought of as coordinates on the worldsheet. Or, another way to look at this is to parameterize the worldsheet, we need to account for proper time *and* the spatial extension of the string. So, the first parameter is once again the proper time τ, and the second parameter, which is related to the length along the string, is denoted by σ:

$$\xi_1 = \tau \qquad \xi_2 = \sigma$$

Coordinates on the worldsheet (τ, σ) are mapped onto space-time by the functions (called the *string coordinates*)

$$X^\mu(\tau, \sigma) \tag{2.12}$$

So time and spatial position on the string are mapped onto the spatial coordinates in $(d + 1)$ dimensional space-time as

$$\{X^0(\tau,\sigma), X^1(\tau,\sigma), \ldots, X^d(\tau,\sigma)\}$$

Now, we need to write down the action for the string which will generalize Eq. (2.8) to our new higher-dimensional world, that is, to the case of the worldsheet. This is done in the following way. Recall that the action of a point particle is proportional to the length of its world-line [Eq. (2.5)]. We just noted that a string sweeps out a two-dimensional worldsheet in space-time. This tells us that if we are going to generalize the notion of the action of a point particle, we might expect that the action of a string is proportional to the surface area of the worldsheet. This is in fact the case. Anticipating that the constant of proportionality will turn out to be the string tension, we can write this action as

$$S = -T \int dA \tag{2.13}$$

where dA is a differential element of area on the worldsheet. To find the form of dA, we start by considering a differential line element ds^2 and introduce coordinates on the worldsheet as $\xi^1 = \tau$ and $\xi^2 = \sigma$. Doing a little algebra we have

$$ds^2 = -\eta_{\mu\nu} dX^\mu dX^\nu$$

$$= -\eta_{\mu\nu} \frac{\partial X^\mu}{\partial \xi^\alpha} \frac{\partial X^\nu}{\partial \xi^\beta} d\xi^\alpha d\xi^\beta$$

This allows us to define an *induced metric* on the worldsheet. This is given by

$$\gamma_{\alpha\beta} = \eta_{\mu\nu} \frac{\partial X^\mu}{\partial \xi^\alpha} \frac{\partial X^\nu}{\partial \xi^\beta} \tag{2.14}$$

This metric determines distances on the worldsheet. We say that this metric is induced because it includes the metric of the background space-time in its definition (we are taking space-time to be flat, so are using $\eta_{\mu\nu}$). That is to say, on the surface of the worldsheet, there is a new measure of distance, but that measure of distance is determined by the background space-time through its metric (which in general, is not $\eta_{\mu\nu}$). Proceeding, we now have

$$ds^2 = \gamma_{\alpha\beta}d\xi^\alpha d\xi^\beta$$

Using the notations

$$\dot{X}^\mu = \frac{\partial X^\mu}{\partial \tau} \qquad X^{\mu\prime} = \frac{\partial X^\mu}{\partial \sigma}$$

We can write the components of the induced metric (for the case of flat space-time) as

$$\gamma_{\tau\tau} = \eta_{\mu\nu}\frac{\partial X^\mu}{\partial \tau}\frac{\partial X^\nu}{\partial \tau} = \dot{X}^2$$

$$\gamma_{\sigma\tau} = \eta_{\mu\nu}\frac{\partial X^\mu}{\partial \sigma}\frac{\partial X^\nu}{\partial \tau} = \dot{X}\cdot X' = \gamma_{\tau\sigma} = \eta_{\mu\nu}\frac{\partial X^\mu}{\partial \tau}\frac{\partial X^\nu}{\partial \sigma} \qquad (2.15)$$

$$\gamma_{\sigma\sigma} = \eta_{\mu\nu}\frac{\partial X^\mu}{\partial \sigma}\frac{\partial X^\nu}{\partial \sigma} = X'^2$$

Using Eq. (2.15), we can write the induced metric as a matrix in (τ, σ) space

$$\gamma_{\alpha\beta} = \begin{pmatrix} \dot{X}^2 & \dot{X}\cdot X' \\ \dot{X}\cdot X' & X'^2 \end{pmatrix} \qquad (2.16)$$

Notice that the determinant of this matrix is given by

$$\gamma = \det\gamma_{\alpha\beta} = \dot{X}^2 X'^2 - (\dot{X}\cdot X'^2) \qquad (2.17)$$

Let's go back to where we started, seeking an expression for the action

$$S = -T\int dA$$

From elementary calculus, we know that in a given space described by a metric $G_{\alpha\beta}$, an element of surface area is written as

$$dA = \sqrt{-\det G_{\alpha\beta}}\, d^2\xi$$

In our case, the metric we need is the induced metric [Eq. (2.15)]. So we take $dA = \sqrt{-\gamma}\, d\tau d\sigma$. If we integrate from some initial proper time τ_i to some final proper time τ_f and over the length of the string (which we'll denote as ℓ) then the action can be written as

$$S = -T \int_{\tau_i}^{\tau_f} d\tau \int_0^\ell d\sigma \sqrt{-\gamma} \tag{2.18}$$

Or explicitly, using Eq. (2.17)

$$S = -T \int_{\tau_i}^{\tau_f} d\tau \int_0^\ell d\sigma \sqrt{\dot{X}^2 X'^2 - (\dot{X} \cdot X'^2)} \tag{2.19}$$

The actions in Eqs. (2.18) and (2.19) are called the *Nambu-Goto action*, which describes the dynamics of a (classical) relativistic string. As the motion of a point particle in space-time serves to minimize the length of the world-line, the motion of a classical string in space-time acts to minimize the surface area of the worldsheet.

Before moving ahead to a quantum theory of strings, we need to find the equations of motion for the string which we can then later quantize.

Equations of Motion for the String

Now that we have the action in place (and hence the lagrangian) we can obtain the equations of motion for the string. We do this using the action principle, which tells us to vary the action and set the result to 0

$$\delta S = 0$$

When computing the variation of the action, we will derive the equations of motion that will be a partial differential equations—meaning that we will need to specify boundary conditions in order to solve them. There are two different types of strings we need to consider when looking at boundary conditions: open strings and closed strings.

If a string is *open*, this means exactly what it says, that the string is a free piece of string with loose ends moving through space-time. The worldsheet in this case is a strip, and by convention we write the endpoint as $\sigma = \pi$. There are two types of boundary conditions that are possible for open strings. The first is called *Neumann* or *free endpoint boundary conditions*. We can write this boundary condition in terms of the lagrangian or the conjugate momentum:

$$\frac{\delta L}{\delta X'^{\mu}}\bigg|_{\sigma=0,\pi} = 0 \quad \text{or} \quad P^{\sigma}_{\mu}\big|_{\sigma=0,\pi} = 0 \quad \text{(Neumann)} \qquad (2.20)$$

Another way to write Neumann boundary conditions for the open string is

$$\frac{\partial X^{\mu}}{\partial \sigma} = 0 \quad \text{when } \sigma = 0, \pi \qquad (2.21)$$

In this case, no momentum can flow off the ends of the string. This indicates that the ends of the string are free to move about space-time. On the other hand, suppose that the ends of the string are fixed instead. In that case we have *Dirichlet* or *fixed point boundary conditions*. These are given by

$$\frac{\delta L}{\delta \dot{X}^{\mu}}\bigg|_{\sigma=0,\pi} = 0 \quad \text{or} \quad \frac{\partial X^{\mu}}{\partial \tau}\bigg|_{\sigma=0,\pi} = 0 \quad \text{(Dirichlet)} \qquad (2.22)$$

A *closed* string is a little loop moving through space-time. In this case, the worldsheet is a cylinder or a tube. The boundary conditions are periodic, described by

$$X^{\mu}(\tau, \sigma) = X^{\mu}(\tau, \sigma + \pi) \qquad (2.23)$$

Now let's write down the conjugate momenta for the string. First recall the lagrangian

$$L = -T\sqrt{(\dot{X} \cdot X')^2 - (\dot{X})^2 (X')^2}$$

The conjugate momentum corresponding to the coordinate σ is

$$\begin{aligned}
P^{\sigma}_{\mu} &= \frac{\partial L}{\partial X'^{\mu}} = \frac{\partial}{\partial X'^{\mu}}\left(-T\sqrt{(\dot{X} \cdot X')^2 - (\dot{X})^2 (X')^2}\right) \\
&= -\frac{T}{2}\left[(\dot{X} \cdot X')^2 - (\dot{X})^2 (X')^2\right]^{-1/2}\left[2(\dot{X} \cdot X')\dot{X}_{\mu} - 2\dot{X}^2 X'_{\mu}\right] \qquad (2.24) \\
&= -T\frac{(\dot{X} \cdot X')\dot{X}_{\mu} - \dot{X}^2 X'_{\mu}}{\sqrt{(\dot{X} \cdot X')^2 - (\dot{X})^2 (X')^2}}
\end{aligned}$$

and we also have a conjugate momentum corresponding to the coordinate τ:

$$
\begin{aligned}
P^{\tau}_{\mu} = \frac{\partial L}{\partial \dot{X}^{\mu}} &= \frac{\partial}{\partial \dot{X}^{\mu}}\left(-T\sqrt{(\dot{X}\cdot X')^2 - (\dot{X})^2(X')^2}\right) \\
&= -\frac{T}{2}\left[(\dot{X}\cdot X')^2 - (\dot{X})^2(X')^2\right]^{-1/2}\left[2(\dot{X}\cdot X')X'_{\mu} - 2X'^2\dot{X}_{\mu}\right] \quad (2.25) \\
&= -T\frac{(\dot{X}\cdot X')X'_{\mu} - X'^2\dot{X}_{\mu}}{\sqrt{(\dot{X}\cdot X')^2 - (\dot{X})^2(X')^2}}
\end{aligned}
$$

Now, let's vary the action to obtain the equations of motion for the string. First, if it's been awhile since you've had field theory, convince yourself that

$$
\delta\dot{X}^{\mu} = \delta\left(\frac{\partial X^{\mu}}{\partial \tau}\right) = \frac{\partial}{\partial \tau}(\delta X^{\mu})
$$

$$
\delta X'^{\mu} = \delta\left(\frac{\partial X^{\mu}}{\partial \sigma}\right) = \frac{\partial}{\partial \sigma}(\delta X^{\mu})
$$

Then, we can vary the action, and using the conjugate momenta we get

$$
\begin{aligned}
\delta S &= -T\int_{\tau_i}^{\tau_f} d\tau \int_0 d\sigma\left[\frac{\partial L}{\partial \dot{X}^{\mu}}\frac{\partial}{\partial \tau}(\delta X^{\mu}) + \frac{\partial L}{\partial X'^{\mu}}\frac{\partial}{\partial \sigma}(\delta X^{\mu})\right] \\
&= -T\int_{\tau_i}^{\tau_f} d\tau \int_0 d\sigma\left[\Pi^{\tau}_{\mu}\frac{\partial}{\partial \tau}(\delta X^{\mu}) + \Pi^{\sigma}_{\mu}\frac{\partial}{\partial \sigma}(\delta X^{\mu})\right]
\end{aligned}
$$

We can rewrite this expression so that we can get terms multiplied by δX^{μ} by using the product rule from calculus. For example,

$$
\frac{\partial}{\partial \tau}\left(P^{\tau}_{\mu}\delta X^{\mu}\right) = P^{\tau}_{\mu}\frac{\partial}{\partial \tau}\delta X^{\mu} + \delta X^{\mu}\frac{\partial P^{\tau}_{\mu}}{\partial \tau}
$$

$$
\Rightarrow P^{\tau}_{\mu}\frac{\partial}{\partial \tau}\delta X^{\mu} = \frac{\partial}{\partial \tau}\left(P^{\tau}_{\mu}\delta X^{\mu}\right) - \delta X^{\mu}\frac{\partial P^{\tau}_{\mu}}{\partial \tau}
$$

Similarly,

$$P_\mu^\sigma \frac{\partial}{\partial \sigma} \delta X^\mu = \frac{\partial}{\partial \sigma}\left(P_\mu^\sigma \delta X^\mu\right) - \delta X^\mu \frac{\partial P_\mu^\sigma}{\partial \sigma}$$

This means that the variation of the action can be written as

$$\delta S = -T\int_{\tau_i}^{\tau_f} d\tau \int_0^\ell d\sigma \left[\frac{\partial}{\partial \tau}\left(P_\mu^\tau \delta X^\mu\right) - \delta X^\mu \frac{\partial P_\mu^\tau}{\partial \tau}\right]$$

$$-T\int_{\tau_i}^{\tau_f} d\tau \int_0^\ell d\sigma \left[\frac{\partial}{\partial \sigma}\left(P_\mu^\sigma \delta X^\mu\right) - \delta X^\mu \frac{\partial P_\mu^\sigma}{\partial \sigma}\right]$$

Recall from classical mechanics that a variation is defined such that variation at the endpoints is 0, that is, at the initial and final times $\delta X^\mu = 0$. In the case of the endpoints of the string, we can apply either Neumann or Dirichlet boundary conditions so we will have to handle each case differently (more on this as we go along). For now, let's take $\delta X^\mu = 0$ for simplicity. This means that we can throw away the terms in the above expression which are integrals of total derivatives:

$$\int_{\tau_i}^{\tau_f} d\tau \frac{\partial}{\partial \tau}\left(P_\mu^\tau \delta X^\mu\right) = 0 \qquad \int_0^\ell d\sigma \frac{\partial}{\partial \sigma}\left(P_\mu^\sigma \delta X^\mu\right) = 0$$

This leaves us with

$$0 = \delta S = T\int_{\tau_i}^{\tau_f} d\tau \int_0^\ell d\sigma \left(\delta X^\mu \frac{\partial P_\mu^\tau}{\partial \tau}\right) + T\int_{\tau_i}^{\tau_f} d\tau \int_0^\ell d\sigma \left(\delta X^\mu \frac{\partial P_\mu^\sigma}{\partial \sigma}\right)$$

$$= T\int_{\tau_i}^{\tau_f} d\tau \int_0^\ell d\sigma \delta X^\mu \left(\frac{\partial P_\mu^\tau}{\partial \tau} + \frac{\partial P_\mu^\sigma}{\partial \sigma}\right)$$

This gives us the equation of motion for the string, derived from the Nambu-Goto action:

$$\frac{\partial P_\mu^\tau}{\partial \tau} + \frac{\partial P_\mu^\sigma}{\partial \sigma} = 0 \qquad\qquad (2.26)$$

The Polyakov Action

Quantization using the Nambu-Goto action is not convenient due to the presence of the square root in the lagrangian. It is possible to write down an equivalent action, equivalent in the sense that it leads to the same equations of motion—that does not have the cumbersome square root. This action goes by the name of the *Polyakov action* or by the more modern term the *string sigma model action*.

Look back to the start of the chapter when we considered the point particle. There too, we ran into a situation where the action had a square root and we dealt with it by introducing an auxiliary field $a(\tau)$. We can use the same procedure here, to rewrite the action for the string in a more convenient form. This is done by introducing an intrinsic metric $h_{\alpha\beta}(\tau, \sigma)$, which acts like the auxiliary field. We use the notation $h_{\alpha\beta}$ because the metric can be written as a matrix. We use the indices to denote rows and columns in this matrix. Then, using the notation $h = \det h_{\alpha\beta}$, the Polyakov action can be written as

$$S_P = -\frac{T}{2} \int d^2\sigma \sqrt{-h}\, h^{\alpha\beta} \partial_\alpha X^\mu \partial_\beta X^\nu \eta_{\mu\nu} \tag{2.27}$$

A historical aside: While Polyakov did important work with this action, it was actually proposed by Brink, Di Vecchia, and Howe and independently by Desser and Zumino. Polyakov got his name attached to it by using it in a path integral quantization of the string. It is also called the *string sigma action*.

Mathematical Aside: The Euler Characteristic

The *Euler characteristic* χ is a number which describes the shape of a topological space. Consider a polyhedron, and let V be the number of vertices, E be the number of edges, and F be the number of faces. Then the Euler characteristic is

$$\chi = V - E + F \tag{2.28}$$

In string theory, we often want to know whether or not two geometric shapes or topologies are similar to one another in a specific way. In particular, we want to know if we can continuously deform one shape into another (imagine working with clay and deforming one shape into another without breaking the clay apart, or introducing or losing any holes). Formally, a *homeomorphism* is a deformation of a geometric object into a new shape by stretching or compressing and being it, without tearing or breaking it. For instance, the quintessential example is a donut and a coffee cup (conveniently paired for police officers). You could use a continuous deformation to transform one into the other or vice versa. So we say that a coffee cup and a donut are homeomorphic. On the other hand, a sphere and a donut are not homomorphic—the donut has a hole but a sphere does not. The bottom line is there is no way to transform the donut into the sphere.

If a geometric shape is homeomorphic to a sphere, then the Euler characteristic is

$$\chi = V - E + F = 2 \tag{2.29}$$

Many shapes have an Euler characteristic which vanishes. Some examples of this include a torus, a möbius strip, and a Klein bottle. Another example is a cylinder, which also has $\chi = 0$ (see Fig. 2.2). Why is this interesting for us? If the worldsheet of a string has a vanishing Euler characteristic, then it is possible to write the auxiliary field $h_{\alpha\beta}$ as a two-dimensional flat space metric. That is, we take [using the choice of coordinates for the worldsheet as (τ, σ)]

$$h_{\alpha\beta} = \begin{pmatrix} -1 & 0 \\ 0 & 1 \end{pmatrix} \tag{2.30}$$

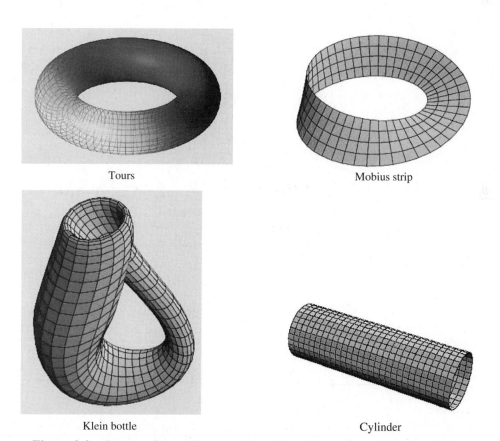

Tours

Mobius strip

Klein bottle

Cylinder

Figure 2.2   Some surfaces with a vanishing Euler characteristic. When the Euler characteristic vanishes, we can define the auxiliary field such that it has a representation of the flat space Minkowski metric.

Now notice that with this choice, $h = \det h_{\alpha\beta} = -1$. We also have

$$h^{\alpha\beta}\partial_\alpha X \cdot \partial_\beta X = -\partial_\tau X \cdot \partial_\tau X + \partial_\sigma X \cdot \partial_\sigma X = -\dot{X}^2 + X'^2$$

In this case, we are able to write the Polyakov action in the remarkably simple form

$$S_P = \frac{T}{2}\int d^2\sigma (\dot{X}^2 - X'^2) \tag{2.31}$$

EXAMPLE 2.3
Find the equations of motion using the Polyakov action as written in Eq. (2.27) when the auxiliary field takes the form of the flat space metric.

SOLUTION
In this case we have

$$S_P = -\frac{T}{2}\int d^2\sigma \sqrt{-h}\, h^{\alpha\beta}\partial_\alpha X^\mu \partial_\beta X^\nu \eta_{\mu\nu}$$

$$= -\frac{T}{2}\int d^2\sigma(-\partial_\tau X \cdot \partial_\tau X + \partial_\sigma X \cdot \partial_\sigma X)$$

$$= -\frac{T}{2}\int d^2\sigma(-\eta_{\mu\nu}\partial_\tau X^\mu \partial_\tau X^\nu + \eta_{\mu\nu}\partial_\sigma X^\mu \partial_\sigma X^\nu)$$

So, we can write the lagrangian as

$$L = -\eta_{\mu\nu}\partial_\tau X^\mu \partial_\tau X^\nu + \eta_{\mu\nu}\partial_\sigma X^\mu \partial_\sigma X^\nu$$

$$= -\eta_{\mu\nu}\dot{X}^\mu \dot{X}^\nu + \eta_{\mu\nu}X'^\mu X'^\nu$$

Therefore,

$$\frac{\partial L}{\partial \dot{X}^\mu} = \frac{\partial}{\partial \dot{X}^\mu}\left(-\eta_{\mu\nu}\dot{X}^\mu \dot{X}^\nu + \eta_{\mu\nu}X'^\mu X'^\nu\right) = -\eta_{\mu\nu}\dot{X}^\nu = -\dot{X}_\mu$$

$$\frac{\partial L}{\partial X'^\mu} = \frac{\partial}{\partial X'^\mu}\left(-\eta_{\mu\nu}\dot{X}^\mu \dot{X}^\nu + \eta_{\mu\nu}X'^\mu X'^\nu\right) = \eta_{\mu\nu}X'^\nu = X'_\mu$$

The Euler-Lagrange equations are

$$\partial_\tau \left(\frac{\partial L}{\partial \dot{X}^\mu} \right) + \partial_\sigma \left(\frac{\partial L}{\partial X'^\mu} \right) = 0 \tag{2.32}$$

Hence, we find that the equations of motion for the relativistic string are

$$\frac{\partial^2 X_\mu}{\partial \tau^2} - \frac{\partial^2 X_\mu}{\partial \sigma^2} = 0 \tag{2.33}$$

Light-Cone Coordinates

It will be convenient to call upon *light-cone coordinates* in string theory. First, let's look at how light-cone coordinates can be defined in Minkowski space-time in general and then consider having them in the context of the worldsheet and the equations of motion of the string. As we will see, this will simplify the way we write the action and the resulting equations of motion.

For simplicity, let's take ordinary (3 + 1) dimensional space-time. The contravariant coordinates are

$$x^\mu = (x^0, x^1, x^2, x^3)$$

where $x^0 = ct$ and $x^1 = x$, $x^2 = y$, $x^3 = z$ say. We form light-cone coordinates by choosing one spatial direction, which in this case we take to be x^1, and forming linear combinations of it with x^0 as follows:

$$x^+ = \frac{x^0 + x^1}{\sqrt{2}} \qquad x^- = \frac{x^0 - x^1}{\sqrt{2}} \tag{2.34}$$

These are two *null* or lightlike coordinates, but you can think of x^+ as a timelike coordinate and x^- as a spacelike coordinate. Hence when we use indices and summations, we will treat + as a "0" index and − as a "1" index. The other coordinates x^2 and x^3 are left alone.

It is easy to derive the inverse relationship using Eq. (2.34). We have

$$x^0 = \frac{x^+ + x^-}{\sqrt{2}} \qquad x^1 = \frac{x^+ - x^-}{\sqrt{2}}$$

Using the Minkowski metric, we have seen that infinitesimal distances in space-time can be defined according to

$$ds^2 = -\eta_{\mu\nu}dx^\mu dx^\nu = (dx^0)^2 - (dx^1)^2 - (dx^2)^2 - (dx^3)^2$$

Since,

$$dx^0 = \frac{dx^+ + dx^-}{\sqrt{2}} \qquad dx^1 = \frac{dx^+ - dx^-}{\sqrt{2}} \tag{2.35}$$

we can rewrite ds^2 in terms of light-cone coordinates as

$$ds^2 = 2dx^+ dx^- - (dx^2)^2 - (dx^3)^2 \tag{2.36}$$

So, we can define distances in terms of a light-cone Minkowski metric

$$\hat{\eta}_{\mu\nu} = \begin{pmatrix} 0 & -1 & 0 & 0 \\ -1 & 0 & 0 & 0 \\ 0 & 0 & 1 & 0 \\ 0 & 0 & 0 & 1 \end{pmatrix} \tag{2.37}$$

Using Eq. (2.37), distances can be written compactly as

$$ds^2 = -\hat{\eta}_{\mu\nu}dx^\mu dx^\nu \tag{2.38}$$

Working with vectors is a simple extension of what we've written for coordinates. That is, define light-cone *components* of a vector v^μ as

$$v^+ = \frac{v^0 + v^1}{\sqrt{2}} \qquad v^- = \frac{v^0 - v^1}{\sqrt{2}} \tag{2.39}$$

Using the metric from Eq. (2.37) the inner product between two vectors can be calculated as

$$v \cdot w = \hat{\eta}_{\mu\nu}v^\mu w^\nu = v^\mu w_\mu = -v^+ w^- - v^- w^+ + \sum_i v^i w^i \tag{2.40}$$

where generally, $i = 1, ..., d-1$. We can apply index raising and lowering to the light-cone components of vectors using a sign change

$$v^+ = -v_- \qquad v^- = -v_+$$

The other components of the vector are left unchanged, that is, $v^i = v_i$.

Now that we've seen how to define light-cone coordinates for space-time, let's see how to define them for the worldsheet and hence for the string. In this case, we define

$$\sigma^+ = \tau + \sigma \qquad \sigma^- = \tau - \sigma \tag{2.41}$$

Now since $d\sigma^+ = d\tau + d\sigma$ and $d\sigma^- = d\tau - d\sigma$, it should be clear that

$$ds^2 = -d\sigma^+ d\sigma^- \tag{2.42}$$

This tells us that we can write the induced metric in Eq. (2.30) indexing a matrix as $(++,+-,-+,--)$ giving

$$h_{\alpha\beta} = \begin{pmatrix} 0 & -1/2 \\ -1/2 & 0 \end{pmatrix} \tag{2.43}$$

You can quickly verify that the determinant is $h = \det h_{\alpha\beta} = -1/4$ and the inverse of Eq. (2.43) is

$$h^{\alpha\beta} = \begin{pmatrix} 0 & -2 \\ -2 & 0 \end{pmatrix}$$

A relationship can also be written down between the derivatives with respect to the coordinates τ, σ and the light-cone coordinates. For notational convenience, we use the relativistic shorthand notation for derivatives

$$\partial_i = \frac{\partial}{\partial x^i}$$

and write

$$\partial_+ = \frac{1}{2}(\partial_\tau + \partial_\sigma) \qquad \partial_- = \frac{1}{2}(\partial_\tau - \partial_\sigma) \tag{2.44}$$

Let's see how the action for the string is written using light-cone coordinates. The Polyakov action, which we reproduce here for your convenience, is

$$S_P = -\frac{T}{2} \int d^2\sigma \sqrt{-h}\, h^{\alpha\beta} \partial_\alpha X^\mu \partial_\beta X^\nu \eta_{\mu\nu}$$

Using Eq. (2.43), we find that

$$\sqrt{-h}\, h^{\alpha\beta} \partial_\alpha X^\mu \partial_\beta X^\nu \eta_{\mu\nu} = -\sqrt{1/4}h^{+-}\partial_+ X^\mu \partial_- X^\nu \eta_{\mu\nu} - \sqrt{1/4}h^{-+}\partial_- X^\mu \partial_+ X^\nu \eta_{\mu\nu}$$

$$= -2\partial_+ X^\mu \partial_- X^\nu \eta_{\mu\nu}$$

Hence, using light-cone coordinates we find the Polyakov action can be written as

$$S_P = T \int d^2\sigma\, \partial_+ X^\mu \partial_- X^\nu \eta_{\mu\nu} \tag{2.45}$$

We can find the equations of motion by varying S_P. We have

$$\delta S_P = \delta T \int d^2\sigma\, \partial_+ X^\mu \partial_- X^\nu \eta_{\mu\nu}$$

$$= T \int d^2\sigma\, \delta\, (\partial_+ X^\mu \partial_- X^\nu) \eta_{\mu\nu}$$

$$= T \int d^2\sigma\, \delta(\partial_+ X^\mu) \partial_- X^\nu \eta_{\mu\nu} + T \int d^2\sigma\, \partial_+ X^\mu \delta(\partial_- X^\nu) \eta_{\mu\nu}$$

The following fact helps us proceed:

$$\delta \frac{\partial X^\mu}{\partial \sigma^\pm} = \frac{\partial(\delta X^\mu)}{\partial \sigma^\pm}$$

Therefore,

$$\delta S_P = T \int d^2\sigma\, \partial_+ (\delta X^\mu) \partial_- X^\nu \eta_{\mu\nu} + T \int d^2\sigma\, \partial_+ X^\mu \, \partial_- (\delta X^\nu) \eta_{\mu\nu}$$

Now integrate by parts to move the derivative away from the δX^μ term. Remember that

$$\int u\, dv = uv - \int v\, du$$

In our case, we get

$$\delta S_P = -T \int d^2\sigma \ (\delta X^\mu)\partial_+\partial_- X^\nu \eta_{\mu\nu} - T \int d^2\sigma \partial_-\partial_+ X^\mu (\delta X^\nu)\eta_{\mu\nu}$$

We've dropped the boundary terms, which must vanish for Neumann boundary conditions in the case of open strings or for the requirement of periodicity for closed strings. Since δX^μ is arbitrary and $\delta S_P = 0$, it must be the case that

$$\partial_+\partial_- X^\mu = 0 \qquad\qquad (2.46)$$

This is the wave equation for relativistic strings using light-cone coordinates.

Solutions of the Wave Equation

In the next chapter, we will consider the hamiltonian and stress-energy tensor and write down conserved charges and currents for the string. Right now, let's focus on finding a solution of the wave equation given in Eq. (2.46).

From elementary mechanics, we know that the solution of a wave equation can be written in terms of a superposition of waves moving to the left on the string and waves moving to the right on the string. If the motion is in one dimension (call it x), then we can write down a solution of the form

$$f(t,\, x) = f_L(x - vt) + f_R(x + vt)$$

We will write the equations of motion for the relativistic string in the same way. We have a solution which is a superposition of left-moving components $X_L^\mu(\tau+\sigma)$ and right-moving components $X_R^\mu(\tau-\sigma)$:

$$X^\mu(\tau,\, \sigma) = X_L^\mu(\tau+\sigma) + X_R^\mu(\tau-\sigma) \qquad\qquad (2.47)$$

You should recall from partial differential equations that the most general solution can be written as an expansion of Fourier modes. Here, we denote these modes as α_k^μ, and write the left-moving and right-moving components as

$$X_L^\mu(\tau,\, \sigma) = \frac{x^\mu}{2} + \frac{\ell_s^2}{2}p^\mu(\tau+\sigma) + i\frac{\ell_s}{\sqrt{2}}\sum_{k\neq 0}\frac{\alpha_k^\mu}{k}e^{-ik(\tau+\sigma)} \qquad\qquad (2.48)$$

$$X_R^\mu(\tau,\, \sigma) = \frac{x^\mu}{2} + \frac{\ell_s^2}{2}\overline{p}^\mu(\tau-\sigma) + i\frac{\ell_s}{\sqrt{2}}\sum_{k\neq 0}\frac{\overline{\alpha}_k^\mu}{k}e^{-ik(\tau-\sigma)} \qquad\qquad (2.49)$$

We have introduced some new terms here. First, we have included the characteristic length of the string which is related to the Regge slope parameter α' and hence to the tension in the string via

$$T = \frac{1}{2\pi\alpha'} \qquad \frac{1}{2}\ell_s^2 = \alpha' \qquad (2.50)$$

Next, notice the coordinate x^μ and momentum p^μ. These are the center of mass coordinate and the total momentum of the string, respectively. The "zeroth"-order Fourier mode is defined in terms of

$$\alpha_0^\mu = \frac{\ell_s}{\sqrt{2}}p^\mu \qquad \bar{\alpha}_0^\mu = \frac{\ell_s}{\sqrt{2}}\bar{p}^\mu \qquad (2.51)$$

What does this tell us physically? The solutions imply that the string can move as a single unit with position and momentum through space-time. In addition, it also has vibrations, which are described by the modes α_k^μ. When you see modes like this, you should think quantization (think in terms of the harmonic oscillator or fields in quantum field theory).

Remember, we are still in the realm of classical physics, even if it's relativistic classical physics. So the solutions of the wave equation X^μ, X_L^μ, and X_R^μ must be *real* functions. This implies that x^μ and p^μ are real (as they must be, given their physical interpretation) and allows us to relate positive and negative modes

$$\alpha_{-k}^\mu = \left(\alpha_k^\mu\right)^* \qquad \bar{\alpha}_{-k}^\mu = \left(\bar{\alpha}_k^\mu\right)^* \qquad (2.52)$$

where * represents the complex conjugate. Now, let's take a look at the solutions of the wave equation with different boundary conditions.

Open Strings with Free Endpoints

Open strings with free endpoints satisfy the Neumann boundary condition that we reproduce here:

$$\frac{\partial X^\mu}{\partial \sigma} = 0 \qquad \text{when } \sigma = 0, \pi$$

Now, looking at Eqs. (2.48) and (2.49) we see that

$$\frac{\partial X_L^\mu}{\partial \sigma} = \frac{\ell_s^2}{2} p^\mu + \frac{\ell_s}{\sqrt{2}} \sum_{k \neq 0} \alpha_k^\mu e^{-ik(\tau+\sigma)}$$

$$\frac{\partial X_R^\mu}{\partial \sigma} = -\frac{\ell_s^2}{2} \overline{p}^\mu - \frac{\ell_s}{\sqrt{2}} \sum_{k \neq 0} \overline{\alpha}_k^\mu e^{-ik(\tau+\sigma)}$$

Summing these as in Eq. (2.47) and setting $\sigma = 0$,

$$\frac{\partial X^\mu}{\partial \sigma} = 0$$

$$\Rightarrow 0 = \frac{\ell_s^2}{2}(p^\mu - \overline{p}^\mu) + \frac{\ell_s}{\sqrt{2}} \sum_{k \neq 0} \left(\alpha_k^\mu - \overline{\alpha}_k^\mu \right) e^{-ik\tau}$$

This tells us that in the case of an open string with free endpoints, it must be the case that

$$p^\mu = \overline{p}^\mu \qquad \text{(string cannot wind around itself)}$$
$$\alpha_k^\mu = \overline{\alpha}_k^\mu \qquad \text{(same modes for left- and right-moving waves)}$$

Physically, this means that for an open string with free endpoints the modes combine to form *standing waves on the string*. Before we move on to our next case, let's consider the other boundary condition, which is imposed at the other end of the string $\sigma = \pi$.

$$0 = \frac{\partial X_L^\mu}{\partial \sigma}\Big|_{\sigma=\pi} + \frac{\partial X_R^\mu}{\partial \sigma}\Big|_{\sigma=\pi}$$

$$= \frac{\ell_s^2}{2} p^\mu + \frac{\ell_s}{\sqrt{2}} \sum_{k \neq 0} \alpha_k^\mu e^{-ik(\tau+\pi)} - \frac{\ell_s^2}{2} p^\mu - \frac{\ell_s}{\sqrt{2}} \sum_{k \neq 0} \alpha_k^\mu e^{-ik(\tau+\pi)}$$

$$= \frac{\ell_s}{\sqrt{2}} \sum_{k \neq 0} \alpha_k^\mu e^{-ik\tau} \left(\frac{e^{-ik\pi} - e^{ik\pi}}{2i} \right)(2i)$$

$$= i\sqrt{2}\,\ell_s \sum_{k \neq 0} \alpha_k^\mu e^{-ik\tau} \sin(k\pi)$$

This can only be true if $\sin k\pi = 0$, which means that k must be an integer. Denoting it by n, a simple exercise shows that Eq. (2.47) can be written as

$$X^{\mu} = x^{\mu} + \ell_{s}^{2} p^{\mu}\tau + i \frac{\ell_{s}}{\sqrt{2}} \sum_{n\neq0} \frac{\alpha_{n}^{\mu}}{n} e^{-in\tau} \cos(n\sigma) \tag{2.53}$$

Closed Strings

In the case of closed strings, the boundary condition becomes one of periodicity, namely

$$X^{\mu}(\tau,\sigma) = X^{\mu}(\tau,\sigma + 2\pi) \tag{2.54}$$

This condition restricts the solutions to those for which the wave number k takes on integral values. Hence,

$$X_{L}^{\mu}(\tau,\sigma) = \frac{x^{\mu}}{2} + \frac{\ell_{s}^{2}}{2} p^{\mu}(\tau+\sigma) + i \frac{\ell_{s}}{\sqrt{2}} \sum_{n\neq0} \frac{\alpha_{n}^{\mu}}{n} e^{-in(\tau+\sigma)} \tag{2.55}$$

$$X_{R}^{\mu}(\tau,\sigma) = \frac{x^{\mu}}{2} + \frac{\ell_{s}^{2}}{2} \overline{p}^{\mu}(\tau-\sigma) + i \frac{\ell_{s}}{\sqrt{2}} \sum_{n\neq0} \frac{\overline{\alpha}_{n}^{\mu}}{n} e^{-in(\tau-\sigma)} \tag{2.56}$$

where n is an integer. In addition, periodicity enforces the condition that

$$p^{\mu} = \overline{p}^{\mu} \tag{2.57}$$

for closed strings. We saw that if this condition was satisfied in the case of open strings, there was no winding of the string permitted. In the case of closed strings, however, the situation is a little bit more involved if we allow for the possibility where the ambient space-time includes a compact extra dimension (then $p^{\mu} = \overline{p}^{\mu}$ does not hold). Then, we can consider, for example, the situation where the closed string is compactified on a circle of radius R.

Using Eq. (2.57), we sum Eqs. (2.55) and (2.56) to obtain the complete solution, focusing on the momentum term–and imposing the periodicity condition of Eq. (2.54). This gives the total solution which can be written as

$$X^{\mu}(\tau,\sigma + 2\pi R) = X^{\mu}(\tau,\sigma) + 2\pi RW \tag{2.58}$$

We call W the *winding number*, which literally tells us how many times the string has wound around the compact dimension (so W must be an integer). As we let $\sigma \to \sigma + 2\pi$, notice that the momentum terms change as

$$\frac{\ell_s^2}{2} p^\mu (\tau + \sigma + \pi) + \frac{\ell_s^2}{2} \bar{p}^\mu (\tau - \sigma - \pi) = \frac{\ell_s^2}{2} p^\mu (\tau + \sigma) + \frac{\ell_s^2}{2} \bar{p}^\mu (\tau - \sigma) + \frac{\pi \ell_s^2}{2} (p^\mu - \bar{p}^\mu)$$

That is, the total solution changes as

$$X^\mu (\tau, \sigma + \pi) = X^\mu (\tau, \sigma) + \frac{\pi \ell_s^2}{2} (p^\mu - \bar{p}^\mu) \tag{2.59}$$

Therefore, we call $(p^\mu - \bar{p}^\mu)$ the *winding contribution*.

Open Strings with Fixed Endpoints

Finally, we consider open strings with fixed endpoints. The boundary condition is

$$\dot{X}^\mu \big|_{\sigma=0} = 0 \tag{2.60}$$

(Dirichlet boundary condition). Using Eqs. (2.48) and (2.49), we have

$$\dot{X}_L^\mu (\tau, \sigma = 0) = \frac{\ell_s^2}{2} p^\mu + \frac{\ell_s}{\sqrt{2}} \sum_{k \neq 0} \alpha_k^\mu e^{-ik\tau} \tag{2.61}$$

$$\dot{X}_R^\mu (\tau, \sigma = 0) = \frac{\ell_s^2}{2} \bar{p}^\mu + \frac{\ell_s}{\sqrt{2}} \sum_{k \neq 0} \bar{\alpha}_k^\mu e^{-ik\tau} \tag{2.62}$$

So, Eq. (2.60) implies that

$$p^\mu + \bar{p}^\mu = 0$$
$$\Rightarrow \bar{p}^\mu = -p^\mu \tag{2.63}$$

and

$$\alpha_k^\mu + \bar{\alpha}_k^\mu = 0 \tag{2.64}$$

If a dimension is noncompact for an open string, then $\bar{p}^\mu = p^\mu$. To simultaneously satisfy Eq. (2.63), the total momentum of the string must vanish. In the next example, we consider the case where both endpoints are fixed.

EXAMPLE 2.4

What is the length of a string that has both endpoints fixed?

SOLUTION

If both endpoints are fixed, then we must also satisfy the boundary condition

$$\dot{X}_L^\mu(\tau, \sigma = \pi) + \dot{X}_R^\mu(\tau, \sigma = \pi) = 0$$

In this case, we have

$$\dot{X}_L^\mu(\tau, \sigma = \pi) = \frac{\ell_s^2}{2} p^\mu + \frac{\ell_s}{\sqrt{2}} \sum_{k \neq 0} \alpha_k^\mu e^{-ik(\tau + \pi)}$$

$$\dot{X}_R^\mu(\tau, \sigma = \pi) = \frac{\ell_s^2}{2} \bar{p}^\mu + \frac{\ell_s}{\sqrt{2}} \sum_{k \neq 0} \bar{\alpha}_k^\mu e^{-ik(\tau - \pi)}$$

The boundary condition can only be satisfied if k is an integer. The overall solution in this case can be written as

$$X^\mu = x^\mu + \ell_s^2 p^\mu \sigma - \sqrt{2} \ell_s \sum_{n \neq 0} \frac{\alpha_n^\mu}{n} e^{-in\tau} \sin(n\sigma) \tag{2.65}$$

Here we applied the conditions in Eqs. (2.63) and (2.64). This expression includes the winding term

$$w = \ell_s^2 p^\mu \tag{2.66}$$

Now let's compute the string coordinates at the endpoints. We have

$$X^\mu(\tau, 0) = x^\mu$$
$$X^\mu(\tau, \pi) = x^\mu + w\pi$$

Hence, the length of the string is

$$X^\mu(\tau, \pi) - X^\mu(\tau, 0) = w\pi$$

Poisson Brackets

In going from ordinary classical mechanics to quantum theory, we follow Dirac and use the correspondence between the Poisson brackets and commutators. In string theory, we consider equal τ Poisson brackets as our starting point when we quantize the modes on the strings. Later, we will discuss hamiltonians and the Virasoro algebra, an important concept in string theory, but for now we simply introduce the important Poisson brackets which will allow us to quantize the string. First we note that

$$\{X^{\mu}(\tau,\sigma),\dot{X}^{\nu}(\tau,\sigma')\} = \frac{1}{T}\delta(\sigma - \sigma')\eta_{\mu\nu} \tag{2.67}$$

or, in terms of momentum

$$\{P^{\mu}(\tau,\sigma),X^{\nu}(\tau,\sigma')\} = \delta(\sigma - \sigma')\eta_{\mu\nu} \tag{2.68}$$

The Fourier expansion can be used to derive Poisson brackets for the modes. These are

$$\left\{\alpha_{m}^{\mu},\,\alpha_{n}^{\nu}\right\} = -im\delta_{m+n,\sigma}\eta_{\mu\nu} \tag{2.69}$$

The Poisson brackets will be the starting point to quantize the theory.

Quiz

1. Consider the lagrangian given in the Eq. (2.10) which includes the auxiliary field. Use the Euler-Lagrange equations to derive the equation of motion.

2. Start with the Nambu-Goto lagrangian $L = \sqrt{\dot{X}^2 X'^2 - (\dot{X} \cdot X'^2)}$ and consider the gauge choice which gives the flat metric $h_{\alpha\beta}$. Using the constraints $\dot{X}^2 + X'^2 = 0$, $\dot{X} \cdot X' = 0$, find the equations of motion.

3. Consider the Polyakov action. A Weyl transformation is one of the form $h_{\alpha\beta} \rightarrow e^{\phi(\tau,\sigma)}h_{\alpha\beta}$ and $\delta X^{\mu} = 0$. Determine the form of the Polyakov action under a Weyl transformation [hint: $h^{\alpha\beta} \rightarrow e^{-\phi(\tau,\sigma)}h^{\alpha\beta}$].

4. Using the Polyakov action, define the energy-momentum tensor on the worldsheet by varying the action with respect to the intrinsic metric

$$T_{\alpha\beta} = -\frac{2}{T}\frac{1}{\sqrt{-h}}\frac{\delta S_P}{\delta h_{\alpha\beta}}$$

Using the fact that $\delta\sqrt{-h} = -\dfrac{1}{2}\sqrt{-h}\, h_{\alpha\beta}\delta h^{\alpha\beta}$ find an expression for the induced metric so that it can be eliminated from the action. This should allow you to recover the Nambu-Goto action.

5. Consider an open string with free endpoints. Find the variation of the center of mass position of the string with τ by calculating

$$\frac{1}{\pi}\int_0^\pi d\sigma X^\mu(\tau,\sigma)$$

6. Find the conserved momentum of the open string with free endpoints by calculating

$$P^\mu = T\int_0^\pi d\sigma \dot{X}^\mu(\tau,\sigma)$$

where the dot represents differentiation with respect to τ.

CHAPTER 3

The Classical String II: Symmetries and Worldsheet Currents

In the last chapter we introduced some basic notions of classical string theory, including the equations of motion and boundary conditions. In this chapter we will expand our discussion of the classical string, discussing symmetries and introducing the energy-momentum tensor and conserved currents. This will finish the groundwork we need for the classical string, and in the next chapter we will quantize the string.

The Energy-Momentum Tensor

Let's quickly review a few things before getting started. Recall that the intrinsic distance on the worldsheet can be determined using the induced metric $h_{\alpha\beta}$. This is given by

$$ds^2 = h_{\alpha\beta}d\sigma^\alpha d\sigma^\beta \tag{3.1}$$

where $\sigma^0 = \tau, \sigma^1 = \sigma$ are the coordinates which parameterize points on the worldsheet. A set of functions $X^\mu(\sigma, \tau)$ describe the shape of the worldsheet and the motion of the string with respect to the background space-time, where $\mu = 0, 1, ..., D-1$ for a D-dimensional space-time. To find the dynamics of the string, we can minimize the Polyakov action [Eq. (2.27)]:

$$S_P = -\frac{T}{2}\int d^2\sigma\sqrt{-\det(h)}\,h^{\alpha\beta}\partial_\alpha X^\mu\partial_\beta X^\nu\,\eta_{\mu\nu} \tag{3.2}$$

Minimizing S_P (by minimizing the area of the worldsheet) gives us the equations of motion for the $X^\mu(\sigma, \tau)$, and hence the dynamics of the string. In the quiz at the end of Chap. 2 in Prob. 4, you were invited to show that the Polyakov and Nambu-Goto actions were equivalent by considering the *energy-momentum* or *stress-energy* tensor $T_{\alpha\beta}$ which is given by

$$T_{\alpha\beta} = -\frac{2}{T}\frac{1}{\sqrt{-h}}\frac{\delta S_P}{\delta h_{\alpha\beta}} \tag{3.3}$$

In this book we'll go mostly by the name energy-momentum tensor. In a nutshell, the energy-momentum tensor describes the density and flux of energy and momentum in space-time. You should be familiar with the basics of what $T_{\alpha\beta}$ is from some exposure to or study of quantum field theory, so we're just going to go with that and describe how it works in string theory. When working out the solution to Prob. 4 in Chap. 2, you should have found that

$$T_{\alpha\beta} = \partial_\alpha X^\mu\partial_\beta X^\nu\eta_{\mu\nu} - \frac{1}{2}h_{\alpha\beta}\left(h^{\rho\sigma}\partial_\rho X^\mu\partial_\sigma X^\nu\eta_{\mu\nu}\right) \tag{3.4}$$

The first property that we will establish for the energy-momentum tensor is that it has zero trace. We can calculate the trace using the induced metric:

$$Tr(T_{\alpha\beta}) = T^\alpha{}_\alpha = h^{\alpha\beta}T_{\alpha\beta}$$

The trace is easy to calculate:

$$h^{\alpha\beta}T_{\alpha\beta} = h^{\alpha\beta}\partial_\alpha X^\mu \partial_\beta X^\nu \eta_{\mu\nu} - \frac{1}{2}h^{\alpha\beta}h_{\alpha\beta}\left(h^{\rho\sigma}\partial_\rho X^\mu \partial_\sigma X^\nu \eta_{\mu\nu}\right)$$

$$= h^{\alpha\beta}\partial_\alpha X^\mu \partial_\beta X^\nu \eta_{\mu\nu} - h^{\rho\sigma}\partial_\rho X^\mu \partial_\sigma X^\nu \eta_{\mu\nu} \quad (\text{because } h^{\alpha\beta}h_{\alpha\beta} = \delta^\alpha_\alpha = 2)$$

$$= h^{\alpha\beta}\partial_\alpha X^\mu \partial_\beta X^\nu \eta_{\mu\nu} - h^{\alpha\beta}\partial_\alpha X^\mu \partial_\beta X^\nu \eta_{\mu\nu} \quad (\text{relabel dummy indices, } \rho \to \alpha, \sigma \to \beta)$$

$$= 0$$

So we've established our first fact to file away about the energy-momentum tensor for the string. It's traceless:

$$T^\alpha_{\ \alpha} = 0 \tag{3.5}$$

In a moment we will learn the physical reason why the energy-momentum tensor is traceless.

Symmetries of the Polyakov Action

In this section we will list some of the symmetries of the Polyakov action (and hence of the bosonic string in Minkowski space-time). There are three symmetry groups of the Polyakov action. These include

- Poincaré transformations
- Reparameterizations of the worldsheet coordinates
- Weyl transformations

The concept of symmetries is so important we will take a momentary digression to discuss the topic before describing the symmetries of the Polyakov action. Symmetries in physics can be *global symmetries* or *local symmetries*. These are defined as follows:

- A global symmetry is one that holds at all points in space-time. The parameters of the transformation will not depend on space-time.

- A local symmetry is one that acts differently at different space-time points. In this case, the parameters of the transformation will be functions of the space-time coordinates.

You should recall from your studies of classical mechanics and quantum field theory that a symmetry in physics leads to a conservation law. The formal statement of this fact is called *Noether's theorem*. Let's quickly review the most famous example of a conserved quantity, the conservation of electric charge.

The electromagnetic field tensor $F_{\mu\nu}$ is defined in terms of the 4-vector potential via

$$F_{\mu\nu} = \partial_\mu A_\nu - \partial_\nu A_\mu$$

The Maxwell equations with source terms are written as

$$\partial_\mu F^{\mu\nu} = J^\nu$$

Now it follows from the definition of $F_{\mu\nu}$ that $\partial_\nu J^\nu = 0$ because

$$\partial_\mu F^{\mu\nu} = \partial_\mu \partial^\mu A^\nu - \partial_\mu \partial^\nu A^\mu \Rightarrow \partial_\nu \partial_\mu F^{\mu\nu}$$

$$= \partial_\nu \partial_\mu \partial^\mu A^\nu - \partial_\nu \partial_\mu \partial^\nu A^\mu$$

$$= \partial_\nu \partial_\mu \partial^\mu A^\nu - \partial_\mu \partial_\nu \partial^\nu A^\mu \quad \text{(partial derivatives commute)}$$

$$= \partial_\nu \partial_\mu \partial^\mu A^\nu - \partial_\nu \partial_\mu \partial^\mu A^\nu \quad \text{(relabel dummy indices)}$$

$$= 0$$

Hence J^μ is a conserved quantity. This is a fact expressed in the famous continuity equation which tells us that

$$\frac{\partial \rho}{\partial t} + \vec{\nabla} \cdot \vec{J} = 0$$

where ρ is the charge density and $\vec{J}$ is the current density. It follows that charge is conserved. The charge Q is of course defined using

$$Q = \int d^3x \, \rho$$

Using the continuity equation and taking the surface of integration S to be at infinity we get

$$\frac{dQ}{dt} = \int d^3x \frac{\partial \rho}{\partial t} = -\int d^3x \, \vec{\nabla} \cdot \vec{J} = -\oint_S \vec{J} \cdot d\vec{A} = 0$$

Hence, charge is conserved.

We demonstrated that charge is conserved starting with the equation of motion for the electromagnetic field. More formally, we can determine what the conserved quantities are and relate them to symmetries by looking at the action S, or more particularly the lagrangian. This is where Noether's theorem comes into play—symmetries in the lagrangian lead to conserved quantities.

You can understand Noether's theorem with a simple one-dimensional example. Consider a particle whose motion is described by a lagrangian $L(q, \dot{q})$ where

$$\dot{q} = \frac{dq}{dt}$$

The momentum of the particle is given by

$$p = \frac{\partial L}{\partial \dot{q}}$$

The Euler-Lagrange equations are the equations of motion for this system:

$$\frac{d}{dt}\frac{\partial L}{\partial \dot{q}} - \frac{\partial L}{\partial q} = 0$$

Now suppose that the lagrangian is invariant under a symmetry. That is, the form of the lagrangian does not change under a one parameter coordinate transformation $t \to s(t)$:

$$q(t) \to q(s)$$

Saying that the lagrangian is invariant under this symmetry means that

$$\frac{d}{ds} L[q(s), \dot{q}(s)] = 0$$

The symmetry of the lagrangian can be written out explicitly using the chain rule as

$$\frac{d}{ds} L[q(s), \dot{q}(s)] = \frac{\partial L}{\partial q}\frac{dq}{ds} + \frac{\partial L}{\partial \dot{q}}\frac{d\dot{q}}{ds} = 0$$

Now let's get to the central idea. Noether's theorem tells us that

$$Q = p\frac{dq}{ds}$$

is a conserved quantity, that is,

$$\frac{dQ}{dt} = 0$$

This is very easy to prove in our one-dimensional example. The calculation is

$$\frac{dQ}{dt} = \frac{d}{dt}\left(p \frac{dq}{ds} \right)$$

$$= \frac{dp}{dt}\frac{dq}{ds} + p\frac{d}{dt}\frac{dq}{ds}$$

$$= \frac{dp}{dt}\frac{dq}{ds} + p\frac{d}{ds}\frac{dq}{dt} \qquad \text{(commutativity of partial derivatives)}$$

$$= \frac{d}{dt}\frac{\partial L}{\partial \dot{q}}\frac{dq}{ds} + \frac{\partial L}{\partial \dot{q}}\frac{d\dot{q}}{ds} \qquad \text{(notation, let } p \to \frac{\partial L}{\partial \dot{q}}, \frac{dq}{dt} \to \dot{q})$$

$$= \frac{d}{dt}\frac{\partial L}{\partial \left(\dfrac{dq}{dt} \right)}\frac{dq}{ds} + \frac{\partial L}{\partial \dot{q}}\frac{d\dot{q}}{ds}$$

$$= \frac{\partial L}{\partial q}\frac{dq}{ds} + \frac{\partial L}{\partial \dot{q}}\frac{d\dot{q}}{ds} = \frac{dL}{ds} = 0 \qquad \text{(symmetry of the lagrangian)}$$

For a field φ^{μ} we define a *Noether current* which is a conserved quantity as

$$j_{\mu}{}^{\alpha} = \frac{\partial L}{\partial\left(\partial_{\alpha}\varphi^{\mu} \right)} \tag{3.6}$$

We are done with our quick and dirty review of symmetries and conserved quantities. Now let's see what symmetries and conserved quantities we can describe for bosonic string theory in D flat space-time dimensions.

POINCARÉ TRANSFORMATIONS

The Poincaré group consists of the following transformations:

- Translations in space-time
- Lorentz transformations

In flat D-dimensional space-time, the Polyakov action is invariant under Poincaré transformations. A space-time translation is a transformation of the form

$$X^{\mu} \to X^{\mu} + b^{\mu} \tag{3.7}$$

where $\delta X^\mu = b^\mu$. An infinitesimal Lorentz transformation is one of the form

$$X^\mu \to X^\mu + \omega^\mu{}_\nu X^\nu \tag{3.8}$$

In this case $\delta X^\mu = \omega^\mu{}_\nu X^\nu$. We can combine translations and infinitesimal Lorentz transformations as

$$\delta X^\mu = \omega^\mu{}_\nu X^\nu + b^\mu \tag{3.9}$$

Under a Poincaré transformation the worldsheet metric transforms as

$$\delta h^{\alpha\beta} = 0 \tag{3.10}$$

The Polyakov action of Eq. (3.2) is invariant under the transformations given in Eqs. (3.9) and (3.10). Invariance under Eq. (3.7) leads to conservation of energy and momentum (energy from time translational invariance and momentum from spatial translation invariance). Invariance of the Polyakov action under Eq. (3.8) leads to conservation of angular momentum.

Recall the definition of a global symmetry and notice that while the transformations in Eqs. (3.7) and (3.8) depend on the coordinates of the embedding space-time (the fields X^μ), they do not depend on the worldsheet coordinates (σ, τ). This means that on the worldsheet, these symmetries are global. Since the symmetry is global on the worldsheet and not over all of space-time, we say that this is a *global internal symmetry*. Put another way, in string theory a global internal symmetry is one that acts on the fields X^μ but not on the two-dimensional space-time of the worldsheet, that is, the parameters of a global internal symmetry group are independent of the worldsheet coordinates (σ, τ).

REPARAMETERIZATIONS

Consider a coordinate transformation that takes $(\sigma, \tau) \to (\sigma', \tau')$, which is a *reparameterization* of the worldsheet (also called a *diffeomorphism*). The metric $h_{\alpha\beta}$ transforms as

$$h_{\alpha\beta} = \frac{\partial \sigma'^\mu}{\partial \sigma^\alpha} \frac{\partial \sigma'^\nu}{\partial \sigma^\beta} h'_{\mu\nu}(\sigma', \tau') \tag{3.11}$$

(note that in this context we are using primes not to denote differentiation, but rather to indicate quantities like the metric in the new coordinate system). Since $\partial/\partial\sigma'^\alpha = (\partial\sigma^\rho/\partial\sigma'^\alpha)(\partial/\partial\sigma^\rho)$ and $X^\mu(\sigma, \tau) \to X'^\mu(\sigma', \tau')$ it follows that

$$h^{\alpha\beta}(\sigma, \tau)\frac{\partial X^\mu}{\partial \sigma^\alpha} \frac{\partial X_\mu}{\partial \sigma^\beta} = h'^{\rho\lambda}(\sigma', \tau')\frac{\partial X'^\mu}{\partial \sigma'^\rho} \frac{\partial X'_\mu}{\partial \sigma'^\lambda}$$

The jacobian for a change in coordinates $\sigma \to \sigma'$ is defined by

$$J = \det\left(\frac{\partial \sigma'^{\alpha}}{\partial \sigma^{\mu}}\right)$$

The jacobian shows up in two places that turn out to cancel themselves to leave the form of the Polyakov action invariant. It shows up when calculating the determinant of the metric as

$$\det(h'_{\alpha\beta}) = J^2 \det(h_{\alpha\beta})$$

You may recall from calculus that it also shows up in the integration measure:

$$d^2\sigma' = J\, d^2\sigma$$

These cancel out in the terms that appear in the Polyakov action [Eq. (3.2)]. That is,

$$d^2\sigma'\sqrt{-\det h'} = d^2\sigma\sqrt{-\det h}$$

Putting all of these results together, we see that a change of worldsheet coordinates (a reparameterization) leaves the Polyakov action invariant. Therefore a reparameterization is a symmetry of the action. Since a reparameterization depends on the worldsheet coordinates (σ, τ), these are *local* symmetries.

WEYL TRANSFORMATIONS

A *Weyl transformation* or *Weyl rescaling* is a conformal transformation of the worldsheet metric (see Chap. 5) of the form:

$$h_{\mu\nu} \to e^{\phi(\sigma,\, \tau)} h_{\mu\nu} \tag{3.12}$$

Since $h^{\alpha\beta}h_{\beta\gamma} = \delta^{\alpha}_{\gamma}$ it follows from Eq. (3.12) that $h^{\mu\nu} \to e^{-\phi(\sigma,\, \tau)}h^{\mu\nu}$. Now we recall two facts about determinants, where we let A, B be $n \times n$ matrices:

$$\det(AB) = \det A \det B$$

$$\det(\alpha A) = \det(\alpha I_n A) = \alpha^n \det A$$

In our case, we are working in two dimensions and so:

$$\det(e^\phi h) = e^{2\phi} \det h$$

This means that we have

$$\sqrt{-\det h}\, h^{\mu\nu} \rightarrow \sqrt{-e^{2\phi}\det h}\, e^{-\phi} h^{\mu\nu} = \sqrt{-\det h}\, h^{\mu\nu}$$

Therefore the Polyakov action is invariant under a Weyl transformation. Since Eq. (3.12) is dependent on the space-time coordinates (σ, τ) of the worldsheet, it is a local symmetry.

Transforming to a Flat Worldsheet Metric

Gauge freedom can be used to simplify the worldsheet metric. What this means is that we can use the symmetries of the action (i.e., utilize the transformations that leave the action and hence the physics unchanged) to write the worldsheet metric in a more convenient form, that is sometimes called the *fiducial metric* $\hat{h}_{\alpha\beta}(\sigma)$. Here we consider a worldsheet with a vanishing Euler characteristic (a cylinder is relevant to our interest).

The worldsheet has only two coordinates (σ, τ), and this means that $h_{\alpha\beta}$ is a 2×2 matrix:

$$h_{\alpha\beta} = \begin{pmatrix} h_{00} & h_{01} \\ h_{10} & h_{11} \end{pmatrix} \tag{3.13}$$

This is going to make things particularly easy for us. We immediately find that there are only three independent components of the worldsheet metric. This is because, in general, the metric tensor $g_{\alpha\beta}$ is symmetric, so that

$$g_{\alpha\beta} = g_{\beta\alpha}$$

The fact that the metric is just 2×2 means that the symmetry requirement fixes the off-diagonal components:

$$h_{01} = h_{10}$$

So, this means that we only have to specify three components of $h_{\alpha\beta}$. The choice can be simplified by using two of the local symmetries of the Polyakov action. From the previous section we recall that these are

- Reparameterization invariance
- Weyl transformations

The first case, reparameterization invariance, that is, a coordinate transformation, can be used to take the metric into a form that is proportional to the two-dimensional flat Minkowski metric $\eta_{\alpha\beta}$ as follows:

$$h_{\alpha\beta} \to e^{\phi(\sigma,\,\tau)}\eta_{\alpha\beta} = e^{\phi(\sigma,\,\tau)}\begin{pmatrix} -1 & 0 \\ 0 & 1 \end{pmatrix} \tag{3.14}$$

This form happens to be particularly useful, because now we can apply a Weyl transformation to get rid of the exponential factor. The end result is that it is possible to use the local symmetries of the Polyakov action to take the worldsheet metric into the flat Minkowski metric:

$$h_{\alpha\beta} \to \eta_{\alpha\beta} = \begin{pmatrix} -1 & 0 \\ 0 & 1 \end{pmatrix} \tag{3.15}$$

This is going to really simplify the situation at hand. First let's write down the Polyakov action [Eq. (2.27)] once again:

$$S_P = -\frac{T}{2}\int d^2\sigma\sqrt{-h}\,h^{\alpha\beta}\partial_\alpha X^\mu \partial_\beta X^\nu \eta_{\mu\nu}$$

The first thing to notice about Eq. (3.15) is that the determinant is just

$$h = \det h_{\alpha\beta} = \det\begin{vmatrix} -1 & 0 \\ 0 & 1 \end{vmatrix} = -1$$

and so

$$\sqrt{-h} = +1$$

Now since

$$h^{\alpha\beta} = \begin{pmatrix} -1 & 0 \\ 0 & 1 \end{pmatrix}$$

we have

$$h^{\alpha\beta}\partial_\alpha X^\mu \partial_\beta X^\nu \eta_{\mu\nu} = h^{\tau\tau}\partial_\tau X^\mu \partial_\tau X^\nu \eta_{\mu\nu} + h^{\sigma\sigma}\partial_\sigma X^\mu \partial_\sigma X^\nu \eta_{\mu\nu}$$

$$= -\partial_\tau X^\mu \partial_\tau X^\nu \eta_{\mu\nu} + \partial_\sigma X^\mu \partial_\sigma X^\nu \eta_{\mu\nu}$$

$$= -\partial_\tau X^\mu \partial_\tau X_\mu + \partial_\sigma X^\mu \partial_\sigma X_\mu$$

Do you remember your quantum field theory? This equation should look familiar—you might recognize the lagrangian (density) for a set of free massless scalar fields. Putting this into the Polyakov action, we see that using its local symmetries has taken it into the very simple form

$$S_P = -\frac{T}{2}\int d^2\sigma \sqrt{-h}\, h^{\alpha\beta}\partial_\alpha X^\mu \partial_\beta X^\nu \eta_{\mu\nu}$$

$$\rightarrow \frac{T}{2}\int d^2\sigma \left(\partial_\tau X^\mu \partial_\tau X_\mu - \partial_\sigma X^\mu \partial_\sigma X_\mu\right) \qquad (3.16)$$

In the following we use abbreviated notation:

$$\frac{\partial X^\mu}{\partial \tau} = \dot{X}^\mu \qquad \frac{\partial X^\mu}{\partial \sigma} = X'^\mu$$

In flat space ($h_{\alpha\beta} = \eta_{\alpha\beta}$) the energy-momentum tensor can be written as

$$T_{\alpha\beta} = \partial_\alpha X^\mu \partial_\beta X_\mu - \frac{1}{2}\eta_{\alpha\beta}\left(\eta^{\lambda\rho}\partial_\lambda X^\mu \partial_\rho X_\mu\right)$$

Let's work out each component. We have

$$T_{\tau\tau} = \partial_\tau X^\mu \partial_\tau X_\mu - \frac{1}{2}\eta_{\tau\tau}\left(\eta^{\tau\tau}\partial_\tau X^\mu \partial_\tau X_\mu + \eta^{\sigma\sigma}\partial_\sigma X^\mu \partial_\sigma X_\mu\right)$$

$$= \partial_\tau X^\mu \partial_\tau X_\mu + \frac{1}{2}\left(-\partial_\tau X^\mu \partial_\tau X_\mu + \partial_\sigma X^\mu \partial_\sigma X_\mu\right)$$

$$= \frac{1}{2}\left(\partial_\tau X^\mu \partial_\tau X_\mu + \partial_\sigma X^\mu \partial_\sigma X_\mu\right) = \frac{1}{2}\left(\dot{X}^\mu \dot{X}_\mu + X'^\mu X'_\mu\right)$$

Next we have

$$T_{\sigma\sigma} = \partial_\sigma X^\mu \partial_\sigma X_\mu - \frac{1}{2}\eta_{\sigma\sigma}\left(\eta^{\tau\tau}\partial_\tau X^\mu \partial_\tau X_\mu + \eta^{\sigma\sigma}\partial_\sigma X^\mu \partial_\sigma X_\mu\right)$$

$$= \partial_\tau X^\mu \partial_\tau X_\mu - \frac{1}{2}\left(-\partial_\tau X^\mu \partial_\tau X_\mu + \partial_\sigma X^\mu \partial_\sigma X_\mu\right)$$

$$= \frac{1}{2}\left(\partial_\tau X^\mu \partial_\tau X_\mu + \partial_\sigma X^\mu \partial_\sigma X_\mu\right) = \frac{1}{2}\left(\dot{X}^\mu \dot{X}_\mu + X'^\mu X'_\mu\right)$$

The off-diagonal terms are

$$T_{\tau\sigma} = \partial_\tau X^\mu \partial_\sigma X_\mu - \frac{1}{2}\eta_{\tau\sigma}\left(\eta^{\tau\tau}\partial_\tau X^\mu \partial_\tau X_\mu + \eta^{\sigma\sigma}\partial_\sigma X^\mu \partial_\sigma X_\mu\right) = \partial_\tau X^\mu \partial_\sigma X_\mu = \dot{X}^\mu X'_\mu$$

$$T_{\sigma\tau} = \partial_\sigma X^\mu \partial_\tau X_\mu - \frac{1}{2}\eta_{\sigma\tau}\left(\eta^{\tau\tau}\partial_\tau X^\mu \partial_\tau X_\mu + \eta^{\sigma\sigma}\partial_\sigma X^\mu \partial_\sigma X_\mu\right) = \partial_\sigma X^\mu \partial_\tau X_\mu = X'^\mu \dot{X}_\mu$$

So we can write the energy-momentum tensor as the matrix

$$T_{\alpha\beta} = \begin{pmatrix} \frac{1}{2}\left(\dot{X}^\mu \dot{X}_\mu + X'^\mu X'_\mu\right) & \dot{X}^\mu X'_\mu \\ \\ X'^\mu \dot{X}_\mu & \frac{1}{2}\left(\dot{X}^\mu \dot{X}_\mu + X'^\mu X'_\mu\right) \end{pmatrix} \tag{3.17}$$

As specified in Eq. (3.5), this energy-momentum tensor has zero trace. This is because

$$Tr(T_{\alpha\beta}) = T^\alpha{}_\alpha = \eta^{\alpha\beta}T_{\alpha\beta} = \eta^{\tau\tau}T_{\tau\tau} + \eta^{\sigma\sigma}T_{\sigma\sigma}$$

$$= -T_{\tau\tau} + T_{\sigma\sigma} = -\frac{1}{2}\left(\dot{X}^\mu \dot{X}_\mu + X'^\mu X'_\mu\right) + \frac{1}{2}\left(\dot{X}^\mu \dot{X}_\mu + X'^\mu X'_\mu\right) = 0$$

Now recall from Chap. 2, Prob. 4, that the energy-momentum tensor is defined using the equation of motion for $h_{\alpha\beta}$. This tells us that the energy-momentum tensor for the string worldsheet is 0, that is,

$$T_{\alpha\beta} = -\frac{2}{T}\frac{1}{\sqrt{-h}}\frac{\delta S_P}{\delta h^{\alpha\beta}} = 0$$

This means that the equations of motion attained from the Polyakov action [Eq. (3.16)] are supplemented by the constraint

$$T_{\alpha\beta} = 0$$

Moreover, in flat space the energy-momentum tensor of the worldsheet is conserved, that is,

$$\partial^{\alpha} T_{\alpha\beta} = 0$$

Conserved Currents from Poincaré Invariance

We can find conserved charges associated with Poincaré invariance which involves charges associated with the global symmetries (translation invariance and Lorentz invariance). The conserved currents (the Noether currents) can be found in the following way. Using a variation of the lagrangian where $\delta X^{\mu} = \varepsilon^{\mu}$, the current J_{μ}^{α} is found from

$$\varepsilon^{\mu} J_{\mu}^{\alpha} = \frac{\partial L}{\partial(\partial_{\alpha} X^{\mu})} \varepsilon^{\mu} \tag{3.18}$$

We will do this in a kind of ad hoc way, using the lagrangian density from the Polyakov action:

$$L_P = -\frac{T}{2}\sqrt{-h}\, h^{\alpha\beta} \partial_{\alpha} X^{\mu} \partial_{\beta} X^{\nu} \eta_{\mu\nu} \tag{3.19}$$

Now consider a translation

$$X^{\mu} \to X^{\mu} + b^{\mu}$$

where b^{μ} is our small parameter. Then

$$L_P \to -\frac{T}{2}\sqrt{-h}\, h^{\alpha\beta} \partial_{\alpha}\left(X^{\mu} + b^{\mu}\right)\partial_{\beta}\left(X^{\nu} + b^{\nu}\right)\eta_{\mu\nu}$$

$$= -\frac{T}{2}\sqrt{-h}\, h^{\alpha\beta}\left(\partial_{\alpha} X^{\mu} + \partial_{\alpha} b^{\mu}\right)\left(\partial_{\beta} X^{\nu} + \partial_{\beta} b^{\nu}\right)\eta_{\mu\nu}$$

$$= -\frac{T}{2}\sqrt{-h}\, h^{\alpha\beta}\left(\partial_{\alpha} X^{\mu}\partial_{\beta} X^{\nu} + \partial_{\alpha} X^{\mu}\partial_{\beta} b^{\nu} + \partial_{\alpha} b^{\mu}\partial_{\beta} X^{\nu} + \partial_{\alpha} b^{\mu}\partial_{\beta} b^{\nu}\right)\eta_{\mu\nu}$$

$$= -\frac{T}{2}\sqrt{-h}\, h^{\alpha\beta}\left(\partial_{\alpha} X^{\mu}\partial_{\beta} X^{\nu} + \partial_{\alpha} X^{\mu}\partial_{\beta} b^{\nu} + \partial_{\alpha} b^{\mu}\partial_{\beta} X^{\nu}\right)\eta_{\mu\nu}$$

Moving to the last line, we dropped the term $\partial_\alpha b^\mu \partial_\beta b^\nu$. This is because we are assuming that b^μ is a small displacement, and so we neglect terms in second order. You will recognize that the first term in the last line is just the original lagrangian (density). So we separate the result as

$$L_P \to -\frac{T}{2}\sqrt{-h}\,h^{\alpha\beta}\left(\partial_\alpha X^\mu \partial_\beta X^\nu + \partial_\alpha X^\mu \partial_\beta b^\nu + \partial_\alpha b^\mu \partial_\beta X^\nu\right)\eta_{\mu\nu}$$

$$=-\frac{T}{2}\sqrt{-h}\,h^{\alpha\beta}\partial_\alpha X^\mu \partial_\beta X^\nu \eta_{\mu\nu} -\frac{T}{2}\sqrt{-h}\,h^{\alpha\beta}\left(\partial_\alpha X^\mu \partial_\beta b^\nu + \partial_\alpha b^\mu \partial_\beta X^\nu\right)\eta_{\mu\nu}$$

$$= L_P + \delta L_P$$

The second term δL_P will be associated with the conserved current. To get it, we want to peel off terms involving b^μ. In order to do this, we will need to get the same indices $\alpha, \beta, \mu,$ and ν on both terms. This is easy because we can exploit the symmetry of the metric. Take a look at the first term. We are going to manipulate it to get the form we want in three steps. First, recalling that repeated indices are dummy indices that we can call what we want, we swap the labels $\mu \leftrightarrow \nu$. Then we exploit the symmetry of the metric to write it the way it originally was, and then we lower an index:

$$-\frac{T}{2}\sqrt{-h}\,h^{\alpha\beta}\left(\partial_\alpha X^\mu \partial_\beta b^\nu\right)\eta_{\mu\nu} = -\frac{T}{2}\sqrt{-h}\,h^{\alpha\beta}\left(\partial_\alpha X^\nu \partial_\beta b^\mu\right)\eta_{\nu\mu} \qquad \text{(relabel dummy indices } \mu \leftrightarrow \nu)$$

$$= -\frac{T}{2}\sqrt{-h}\,h^{\alpha\beta}\left(\partial_\alpha X^\nu \partial_\beta b^\mu\right)\eta_{\mu\nu} \qquad \text{(symmetry of the metric } \eta_{\mu\nu} = \eta_{\nu\mu})$$

$$= -\frac{T}{2}\sqrt{-h}\,h^{\alpha\beta}\left(\partial_\alpha X_\mu \partial_\beta b^\mu\right) \qquad \text{(lower an index)}$$

So now

$$\delta L_P = -\frac{T}{2}\sqrt{-h}\,h^{\alpha\beta}\left(\partial_\alpha X_\mu \partial_\beta b^\mu\right) -\frac{T}{2}\sqrt{-h}\,h^{\alpha\beta}\left(\partial_\alpha b^\mu \partial_\beta X^\nu\right)\eta_{\mu\nu}$$

Now we work on the second term, in two steps. First we lower an index, and then we swap the labels used for the dummy indices $\alpha \leftrightarrow \beta$ and again exploit the symmetry of the metric, but this time it's the worldsheet metric we are talking about:

$$
\begin{aligned}
\delta L_P &= -\frac{T}{2}\sqrt{-h}\,h^{\alpha\beta}\left(\partial_\alpha X_\mu \partial_\beta b^\mu\right) - \frac{T}{2}\sqrt{-h}\,h^{\alpha\beta}\left(\partial_\alpha b^\mu \partial_\beta X^\nu\right)\eta_{\mu\nu} \\
&= -\frac{T}{2}\sqrt{-h}\,h^{\alpha\beta}\left(\partial_\alpha X_\mu \partial_\beta b^\mu\right) - \frac{T}{2}\sqrt{-h}\,h^{\alpha\beta}\left(\partial_\alpha b^\mu \partial_\beta X_\mu\right) \\
&= -\frac{T}{2}\sqrt{-h}\,h^{\alpha\beta}\left(\partial_\alpha X_\mu \partial_\beta b^\mu\right) - \frac{T}{2}\sqrt{-h}\,h^{\beta\alpha}\left(\partial_\beta b^\mu \partial_\alpha X_\mu\right) \\
&= -\frac{T}{2}\sqrt{-h}\,h^{\alpha\beta}\left(\partial_\alpha X_\mu \partial_\beta b^\mu\right) - \frac{T}{2}\sqrt{-h}\,h^{\alpha\beta}\left(\partial_\beta b^\mu \partial_\alpha X_\mu\right) \\
&= -\frac{T}{2}\sqrt{-h}\,h^{\alpha\beta}\left(\partial_\alpha X_\mu \partial_\beta b^\mu\right) - \frac{T}{2}\sqrt{-h}\,h^{\alpha\beta}\left(\partial_\alpha X_\mu \partial_\beta b^\mu\right) \\
&= -T\sqrt{-h}\,h^{\alpha\beta}\left(\partial_\alpha X_\mu\right)\partial_\beta b^\mu
\end{aligned}
$$

Now notice we have a term multiplied by $\partial_\beta b^\mu$, which is the small parameter we used to vary X^μ. The rest of this expression is the conserved current we're looking for:

$$
P_\mu^\beta = -T\sqrt{-h}\,h^{\alpha\beta}(\partial_\alpha X_\mu) \tag{3.20}
$$

If we use reparameterization and Weyl invariance to take $h_{\alpha\beta} \to \eta_{\alpha\beta}$ then we have

$$
P_\mu^\beta = -T\partial_\alpha X_\mu
$$
$$
\Rightarrow P_\mu^\tau = -T\partial_\tau X_\mu \qquad P_\mu^\sigma = -T\partial_\sigma X_\mu
$$

The conservation equation for the current is

$$
\partial_\alpha P_\mu^\alpha = 0 \tag{3.21}
$$

Dropping the string tension T and ignoring the minus sign we see that the conservation equation for the current becomes the equation of motion for the string worldsheet:

$$
\partial_\tau P_\mu^\tau + \partial_\sigma P_\mu^\sigma = 0 \tag{3.22}
$$

P_τ^μ has an immediate physical interpretation. It is the *momentum density* of the string. We integrate along the length of the string fixing τ to get the total momentum carried by the string, which we label p_μ :

$$p_\mu = \int_0^{\sigma_1} d\sigma P_\mu^\tau \tag{3.23}$$

The other conserved current associated with the global symmetries of the action comes from invariance under Lorentz transformations. In this case

$$\delta X^\mu = \omega^\mu{}_\nu X^\nu$$

We can show that the lagrangian in Eq. (3.19) is invariant under a Lorentz transformation in the following way:

$$\delta\left(\partial_\alpha X^\mu \partial_\beta X^\nu \eta_{\mu\nu}\right) = \delta\left(\partial_\alpha X^\mu\right)\partial_\beta X^\nu \eta_{\mu\nu} + \partial_\alpha X^\mu \delta\left(\partial_\beta X^\nu\right)\eta_{\mu\nu}$$

$$= \partial_\alpha\left(\delta X^\mu\right)\partial_\beta X^\nu \eta_{\mu\nu} + \partial_\alpha X^\mu \partial_\beta\left(\delta X^\nu\right)\eta_{\mu\nu}$$

$$= \partial_\alpha\left(\omega^\mu{}_\rho X^\rho\right)\partial_\beta X^\nu \eta_{\mu\nu} + \partial_\alpha X^\mu \partial_\beta\left(\omega^\nu{}_\lambda X^\lambda\right)\eta_{\mu\nu}$$

$$= \omega^\mu{}_\rho \partial_\alpha X^\rho \partial_\beta X^\nu \eta_{\mu\nu} + \omega^\nu{}_\lambda \partial_\alpha X^\mu \partial_\beta X^\lambda \eta_{\mu\nu}$$

$$= \omega_{\nu\rho}\partial_\alpha X^\rho \partial_\beta X^\nu + \omega_{\mu\lambda}\partial_\alpha X^\mu \partial_\beta X^\lambda \quad \text{(lower indices with } \eta_{\mu\nu}\text{)}$$

$$= \omega_{\nu\rho}\partial_\alpha X^\rho \partial_\beta X^\nu + \omega_{\rho\lambda}\partial_\alpha X^\rho \partial_\beta X^\lambda \quad \text{(relabel dummy indices } \mu \to \rho\text{)}$$

$$= \omega_{\nu\rho}\partial_\alpha X^\rho \partial_\beta X^\nu + \omega_{\rho\nu}\partial_\alpha X^\rho \partial_\beta X^\nu \quad \text{(relabel dummy indices } \lambda \to \nu\text{)}$$

$$= \omega_{\nu\rho}\partial_\alpha X^\rho \partial_\beta X^\nu + \omega_{\nu\rho}\partial_\alpha X^\rho \partial_\beta X^\nu = 0 \quad \text{(antisymmetry } \omega_{\alpha\beta} = -\omega_{\beta\alpha}\text{)}$$

So the lagrangian is invariant under a Lorentz transformation, but what are the currents? This is easy to find, since

$$L_P = -\frac{T}{2}\sqrt{-h}\, h^{\alpha\beta}\partial_\alpha X^\mu \partial_\beta X^\nu \eta_{\mu\nu}$$

$$\Rightarrow \frac{\partial L_P}{\partial\left(\partial_\alpha X^\mu\right)} = -\frac{T}{2}\sqrt{-h}\, h^{\alpha\beta}\partial_\beta X^\nu \eta_{\mu\nu} = -\frac{T}{2}\sqrt{-h}\, h^{\alpha\beta} P_\beta^\mu$$

Using $\varepsilon^{\mu\nu}J_{\mu\nu}^\alpha = [\partial L_P / \partial(\partial_\alpha X^\mu)]\delta X^\mu$ where $\delta X^\mu = \omega^\mu{}_\nu X^\nu$ together with the antisymmetry of $\omega^\mu{}_\nu$, we conclude that the Lorentz current is

$$J_\alpha^{\mu\nu} = T\left(X^\mu P_\alpha^\nu - X^\nu P_\alpha^\mu\right) \tag{3.24}$$

The Hamiltonian

We have introduced the energy-momentum tensor and looked at some conserved currents that arise due to symmetries of the lagrangian. The next major piece of the dynamics puzzle is the *hamiltonian* which governs the time evolution of the worldsheet. It can be written down simply using formulas from classical mechanics:

$$H = \int_0^{\sigma_1} d\sigma \left(\dot{X}_\mu P_\tau^\mu - L_P \right) = \frac{T}{2} \int_0^{\sigma_1} d\sigma \left(\dot{X}^2 + X'^2 \right) \tag{3.25}$$

Summary

In this chapter we have extended our classical analysis of the string. We did this by introducing the energy-momentum tensor and by describing the symmetries of the Polyakov action. Then we derived conserved currents of the worldsheet, and wrote down the hamiltonian. In the next chapter, we will conclude our classical description of the string by writing down mode expansions of the hamiltonian and energy-momentum tensor, and describing the Virasoro algebra. After writing down a mass formula for the string, we will proceed to quantize the theory.

Quiz

1. Let $\sigma^\alpha \to \sigma'^\alpha + \varepsilon^\alpha(\sigma)$ be an infinitesimal reparameterization. Considering only terms that are first order in ε^α, find the variation of the worldsheet metric $h^{\alpha\beta}$.

2. Assuming that you can move the derivative $d/d\tau$ inside the integral in Eq. (3.23), explain conservation of momentum for the cases of the open and closed string.

3. The energy-momentum tensor has zero trace. Show that this is a consequence of Weyl invariance.

4. Consider light-cone coordinates and derive the Virasoro constraints for the energy-momentum tensor.

5. Consider the Lorentz current $J_\alpha^{\mu\nu}$ in Eq. (3.24). What is the equation that describes that the current is conserved? What are the conserved charges and what do they describe?

CHAPTER 4

String Quantization

At this point we have the classical physics of the string in place. The next step toward a quantum theory is, of course, to quantize the string. We will start off by looking at a quantum theory of a single string. From quantum field theory you recall that this procedure is known as *first quantization*. This is opposed to the procedure called *second quantization,* which is a viewpoint of quantizing fields (see *Quantum Field Theory Demystified* for a review). The difference in the two approaches will be on how we view the $X^\mu(\sigma, \tau)$. If we take them to be fields, then the quantization procedure is second quantization. If we take them to be space-time coordinates, as we have so far, then the process is first quantization. This is the procedure we will apply in this chapter. There are several different approaches to quantization of the string, each with their own accompanying difficulties and problems. The main approaches used are called *covariant quantization, light-cone quantization,* and *BRST quantization.* We consider the first two approaches here, and will discuss BRST quantization later.

Covariant Quantization

The procedure known as covariant quantization will be familiar to you from your studies of ordinary quantum mechanics. In a nutshell, this is the imposition of commutation relations on position and momentum. So, using this procedure we continue with the notion that $X^{\mu}(\sigma, \tau)$ are space-time coordinates, but we need to say what we are taking as the momentum. This can be done in the standard way using lagrangian dynamics. Let $\pi^{\mu}(\sigma, \tau)$ be the momentum carried by the string. Given a lagrangian density L we can calculate the momentum from the $X^{\mu}(\sigma, \tau)$ using

$$\pi_{\mu}(\sigma, \tau) = \frac{\partial L}{\partial\left(\partial_{\tau} X^{\mu}\right)}$$

This is easy to calculate using the Polyakov action as written in Eq. (2.31):

$$S_P = \frac{T}{2}\int d^2\sigma\left(\partial_{\tau} X^{\mu}\partial_{\tau} X_{\mu} - \partial_{\sigma} X^{\mu}\partial_{\sigma} X_{\mu}\right)$$

$$\Rightarrow L = \frac{T}{2}\left(\partial_{\tau} X^{\mu}\partial_{\tau} X_{\mu} - \partial_{\sigma} X^{\mu}\partial_{\sigma} X_{\mu}\right)$$

So we see that the conjugate momentum is

$$\pi_{\mu}(\sigma, \tau) = \frac{\partial L}{\partial\left(\partial_{\tau} X^{\mu}\right)} = T\partial_{\tau} X_{\mu}$$

With this definition in hand, we are in a position to quantize the theory. To do this we take the approach used in quantum field theory, namely, impose *equal time* commutation relations on the position and momenta. In ordinary quantum mechanics the position and momentum coordinates satisfy

$$[x, p_x] = [y, p_y] = [z, p_z] = i$$
$$[x, x] = [y, y] = [z, z] = [x, y] = [x, z] = [y, z] = 0$$
$$[p_x, p_x] = [p_y, p_y] = [p_z, p_z] = [p_x, p_y] = [p_x, p_z] = [p_y, p_z] = 0$$

where we have set $\hbar = 1$. If we denote the coordinates as x_i where $i = 1, 2, 3$ then these relations can be written compactly as

$$[x_i, p_j] = i\delta_{ij}$$
$$[x_i, x_j] = [p_i, p_j] = 0$$

Here, δ_{ij} is the Kronecker delta. Now we apply these relations to the string, with two crucial differences. First, we have a relativistic theory and hence we let $\delta_{ij} \to \eta_{\mu\nu}$. Moreover, we expect that position and momentum will commute when they are taken to be at different spatial locations of the string. That is, let σ and σ' be two different locations on the string. Since we are taking equal time commutation relations, we let the time coordinate be τ for both position and momentum. Then

$$[X^{\mu}(\sigma, \tau), \pi^{\nu}(\sigma', \tau)] = 0 \qquad \text{for } \sigma \neq \sigma'$$

Now, to ensure that the position and momentum don't commute at the same spatial location, we use the Dirac delta function $\delta(\sigma - \sigma')$. So the equal time commutation relations are:

$$\left[X^{\mu}(\sigma, \tau), \pi^{\nu}(\sigma', \tau)\right] = i\eta^{\mu\nu}\delta(\sigma - \sigma')$$
$$\left[X^{\mu}(\sigma, \tau), X^{\nu}(\sigma', \tau)\right] = \left[\pi^{\mu}(\sigma, \tau), \pi^{\nu}(\sigma', \tau)\right] = 0 \tag{4.1}$$

To summarize, remember that

- τ is the same for $X^{\mu}(\sigma, \tau)$ and $\pi^{\nu}(\sigma', \tau)$.

- The presence of the Dirac delta function $\delta(\sigma - \sigma')$ ensures that the coordinates and momenta do commute at different points σ along the string—the commutation relations are only nonzero when position and momentum are evaluated at the same point on the string.

- The presence of $\eta^{\mu\nu}$ comes from the fact we have a relativistic theory.

Ultimately, we want to write down commutation relations for the *modes* of the string. This can be done most easily by transitioning to the light-cone coordinates $\sigma_{+} = \tau + \sigma$, $\sigma_{-} = \tau - \sigma$. First, using $\partial_{+} = 1/2(\partial_{\tau} + \partial_{\sigma})$ and $\partial_{-} = 1/2(\partial_{\tau} - \partial_{\sigma})$, we can invert to write

$$\partial_{+} + \partial_{-} = \partial_{\tau} \qquad \partial_{+} - \partial_{-} = \partial_{\sigma} \tag{4.2}$$

In the case of $X^{\nu}(\sigma', \tau)$ and $\pi^{\nu}(\sigma', \tau)$, we will have $\partial'_{+} = 1/2 \, (\partial_{\tau} + \partial_{\sigma'})$ and $\partial'_{-} = 1/2 \, (\partial_{\tau} - \partial_{\sigma'})$. Using Eq. (4.2), we can write the commutation relation $[X^{\mu}(\sigma, \tau), \pi^{\nu}(\sigma', \tau)] = i\eta^{\mu\nu}\delta(\sigma - \sigma')$ in a new way:

$$i\eta^{\mu\nu}\delta(\sigma - \sigma') = [X^{\mu}(\sigma, \tau), \pi^{\nu}(\sigma', \tau)] = [X^{\mu}(\sigma, \tau), T\partial_{\tau}X^{\nu}(\sigma', \tau)]$$
$$= T[X^{\mu}(\sigma, \tau), (\partial'_{+} + \partial'_{-})X^{\nu}(\sigma', \tau)]$$
$$= T[X^{\mu}(\sigma, \tau), \partial'_{+}X^{\nu}(\sigma', \tau)] + T[X^{\mu}(\sigma, \tau), \partial'_{-}X^{\nu}(\sigma', \tau)]$$

Now we compute the derivative of $[X^{\mu}(\sigma, \tau), \pi^{\nu}(\sigma', \tau)] = i\eta^{\mu\nu}\delta(\sigma - \sigma')$ with respect to σ. Since $X^{\nu}(\sigma', \tau)$ is not a function of σ, only $X^{\mu}(\sigma, \tau)$ is affected. Proceeding and again using Eq. (4.2) we find

$$i\eta^{\mu\nu}\frac{\partial}{\partial\sigma}\delta(\sigma - \sigma') = T[\partial_{\sigma}X^{\mu}(\sigma, \tau), \partial'_{+}X^{\nu}(\sigma', \tau)]$$

$$+ T[\partial_{\sigma}X^{\mu}(\sigma, \tau), \partial'_{-}X^{\nu}(\sigma', \tau)]$$

$$= T[(\partial_{+} - \partial_{-})X^{\mu}(\sigma, \tau), \partial'_{+}X^{\nu}(\sigma', \tau)]$$

$$+ T[(\partial_{+} - \partial_{-})X^{\mu}(\sigma, \tau), \partial'_{-}X^{\nu}(\sigma', \tau)]$$

$$= T[\partial_{+}X^{\mu}(\sigma, \tau), \partial'_{+}X^{\nu}(\sigma', \tau)] - T[\partial_{-}X^{\mu}(\sigma, \tau), \partial'_{+}X^{\nu}(\sigma', \tau)]$$

$$+ T[\partial_{+}X^{\mu}(\sigma, \tau), \partial'_{-}X^{\nu}(\sigma', \tau)] - T[\partial_{-}X^{\mu}(\sigma, \tau), \partial'_{-}X^{\nu}(\sigma', \tau)]$$

Now we can utilize the commutation relation for the conjugate momenta in Eq. (4.1). We have

$$0 = [\pi^{\mu}(\sigma, \tau), \pi^{\nu}(\sigma', \tau)] = [T\partial_{\tau}X^{\mu}(\sigma, \tau), T\partial_{\tau}X^{\nu}(\sigma', \tau)]$$

$$= T^{2}[\partial_{\tau}X^{\mu}(\sigma, \tau), \partial_{\tau}X^{\nu}(\sigma', \tau)]$$

$$= T[\partial_{\tau}X^{\mu}(\sigma, \tau), \partial_{\tau}X^{\nu}(\sigma', \tau)] \qquad \text{(Since it equals zero, we divide}$$
$$\text{by } T \text{ for later convenience)}$$

$$= T[(\partial_{+} + \partial_{-})X^{\mu}(\sigma, \tau), (\partial'_{+} + \partial'_{-})X^{\nu}(\sigma', \tau)]$$

$$= T[\partial_{+}X^{\mu}(\sigma, \tau), \partial'_{+}X^{\nu}(\sigma', \tau)] + T[\partial_{+}X^{\mu}(\sigma, \tau), \partial'_{-}X^{\nu}(\sigma', \tau)]$$

$$+ T[\partial_{-}X^{\mu}(\sigma, \tau), \partial'_{+}X^{\nu}(\sigma', \tau)] + T[\partial_{-}X^{\mu}(\sigma, \tau), \partial'_{-}X^{\nu}(\sigma', \tau)]$$

Now since $[\pi^{\mu}(\sigma, \tau), \pi^{\nu}(\sigma', \tau)] = 0$, let's form the sum $[\pi^{\mu}(\sigma, \tau), \pi^{\nu}(\sigma', \tau)] + i\eta^{\mu\nu}(\partial/\partial\sigma)\delta(\sigma - \sigma')$. We obtain

$$[\pi^{\mu}(\sigma, \tau), \pi^{\nu}(\sigma', \tau)] + i\eta^{\mu\nu}\frac{\partial}{\partial\sigma}\delta(\sigma - \sigma')$$

$$= T[\partial_{+}X^{\mu}(\sigma, \tau), \partial'_{+}X^{\nu}(\sigma', \tau)] + T[\partial_{+}X^{\mu}(\sigma, \tau), \partial'_{-}X^{\nu}(\sigma', \tau)]$$

$$+ T[\partial_{-}X^{\mu}(\sigma, \tau), \partial'_{+}X^{\nu}(\sigma', \tau)] + T[\partial_{-}X^{\mu}(\sigma, \tau), \partial'_{-}X^{\nu}(\sigma', \tau)]$$

$$+ T[\partial_{+}X^{\mu}(\sigma, \tau), \partial'_{+}X^{\nu}(\sigma', \tau)] - T[\partial_{-}X^{\mu}(\sigma, \tau), \partial'_{+}X^{\nu}(\sigma', \tau)]$$

$$+ T[\partial_{+}X^{\mu}(\sigma, \tau), \partial'_{-}X^{\nu}(\sigma', \tau)] - T[\partial_{-}X^{\mu}(\sigma, \tau), \partial'_{-}X^{\nu}(\sigma', \tau)]$$

That is,

$$i\eta^{\mu\nu}\frac{\partial}{\partial\sigma}\delta(\sigma-\sigma')=2T[\partial_+X^\mu(\sigma,\tau),\partial'_+X^\nu(\sigma',\tau)]+2T[\partial_+X^\mu(\sigma,\tau),\partial'_-X^\nu(\sigma',\tau)]$$

It is also straightforward to show that

$$[\pi^\mu(\sigma,\tau),\pi^\nu(\sigma',\tau)]-i\eta^{\mu\nu}\frac{\partial}{\partial\sigma}\delta(\sigma-\sigma')$$

$$=2T[\partial_-X^\mu(\sigma,\tau),\partial'_-X^\nu(\sigma',\tau)]+2T[\partial_-X^\mu(\sigma,\tau),\partial'_+X^\nu(\sigma',\tau)]$$

In the chapter quiz, you will have the opportunity to show that $[\partial_+X^\mu(\sigma,\tau),\partial_-X^\nu(\sigma,\tau)]=[\partial_-X^\mu(\sigma,\tau),\partial_+X^\nu(\sigma',\tau)]=0$. Therefore the commutation relations are

$$[\partial_+X^\mu(\sigma,\tau),\partial'_+X^\nu(\sigma',\tau)]=\frac{i\eta^{\mu\nu}}{2T}\frac{\partial}{\partial\sigma}\delta(\sigma-\sigma') \tag{4.3}$$

$$[\partial_-X^\mu(\sigma,\tau),\partial'_-X^\nu(\sigma',\tau)]=-\frac{i\eta^{\mu\nu}}{2T}\frac{\partial}{\partial\sigma}\delta(\sigma-\sigma') \tag{4.4}$$

$$[\partial_+X^\mu(\sigma,\tau),\partial'_-X^\nu(\sigma',\tau)]=[\partial_-X^\mu(\sigma,\tau),\partial'_+X^\nu(\sigma',\tau)]=0 \tag{4.5}$$

Using Eqs. (4.3) to (4.5), deriving commutation relations for the modes, which will help us get to the quantum physics, will be a much simpler matter. In order to derive the commutation relations, we will need the following expression for the Dirac delta function:

$$\delta(x)=\frac{1}{2z}\sum_{\kappa=-\infty}^{\infty}e^{ikx}$$

Recalling Eq. (2.47), which told us that we can write the equations for the string in terms of left-moving and right-moving modes:

$$X^\mu(\sigma,\tau)=X_L^\mu(\sigma,\tau)+X_R^\mu(\sigma,\tau)$$

Notice the left-moving modes are functions of σ_+ only, while the right-moving modes are functions of σ_- only, so that

$$\partial_+X^\mu(\sigma,\tau)=\partial_+X_L^\mu(\sigma,\tau)\qquad\partial_-X^\mu(\sigma,\tau)=\partial_-X_R^\mu(\sigma,\tau)$$

COMMUTATION RELATIONS FOR THE CLOSED STRING

We will derive the commutation relation [Eq. (4.3)] for the modes explicitly, and simply state the results in the other cases. Hence we write down the left-moving modes [Eq. (2.55)] which are functions of σ_+. The left-moving modes are restated here in the case of the closed string

$$X_L^\mu(\sigma, \tau) = \frac{x^\mu}{2} + \frac{\ell_s^2}{2} p^\mu(\tau + \sigma) + \frac{i\ell_s}{\sqrt{2}} \sum_{m \neq 0} \frac{\alpha_m^\mu}{m} e^{-im(\tau+\sigma)} \tag{4.6}$$

Now since $\sigma_+ = \tau + \sigma$, the derivative is

$$\partial_+ X_L^\mu(\sigma, \tau) = \frac{\ell_s^2}{2} p^\mu + \frac{\ell_s}{\sqrt{2}} \sum_{m \neq 0} \alpha_m^\mu e^{-im(\tau+\sigma)} = \frac{\ell_s}{\sqrt{2}} \sum_{m=-\infty}^{\infty} \alpha_m^\mu e^{-im(\tau+\sigma)} \tag{4.7}$$

To get the last step, we used $\alpha_0^\mu = (\ell_s/\sqrt{2})p^\mu$. Now let's calculate the left-hand side of Eq. (4.3). We have

$$[\partial_+ X^\mu(\sigma, \tau), \partial'_+ X^\nu(\sigma', \tau)] = \left[\frac{\ell_s}{\sqrt{2}} \sum_{m=-\infty}^{\infty} \alpha_m^\mu e^{-im(\tau+\sigma)}, \frac{\ell_s}{\sqrt{2}} \sum_{m=-\infty}^{\infty} \alpha_m^\nu e^{-im(\tau+\sigma')} \right]$$

$$= \frac{\ell_s^2}{2} \sum_{m,n=-\infty}^{\infty} e^{-i(m+n)\tau} e^{-i(m\sigma+n\sigma')} \left[\alpha_m^\mu, \alpha_m^\nu \right]$$

But, this must be proportional to $(\partial/\partial\sigma)\delta(\sigma - \sigma')$, and furthermore, the term on the right-hand side of Eq. (4.3) does not depend on τ. So, we have to remove the τ dependence. We can do so by noting that

$$e^{-i(m+n)\tau} \to 1 \qquad \text{when} \qquad m = -n$$

We will be able to enforce this condition by introducing the Kronecker delta term $\delta_{m+n,0}$, which is 1 when $m = -n$ and 0 otherwise. Furthermore, we take note of the following expression for the Dirac delta function:

$$\delta(\sigma - \sigma') = \frac{1}{2\pi} \sum_{m=-\infty}^{\infty} e^{-im(\sigma-\sigma')} \tag{4.8}$$

Notice that

$$\frac{\partial}{\partial\sigma} \delta(\sigma - \sigma') = -\frac{i}{2\pi} \sum_{m=-\infty}^{\infty} m e^{-im(\sigma-\sigma')}$$

Using Eqs. (4.3) and (4.8) together with our previous result, we have

$$\frac{\ell_s^2}{2} \sum_{m,n=-\infty}^{\infty} e^{-i(m+n)\tau} e^{-i(m\sigma+n\sigma')} [\alpha_m^\mu, \alpha_m^\nu] = \frac{i\eta^{\mu\nu}}{2T} \frac{\partial}{\partial\sigma} \delta(\sigma-\sigma')$$

$$= \frac{i\eta^{\mu\nu}}{2T} \left(-\frac{i}{2\pi} \sum_{m=-\infty}^{\infty} m e^{-im(\sigma-\sigma')} \right)$$

$$= \frac{\eta^{\mu\nu}}{2} \frac{1}{2\pi T} \sum_{m=-\infty}^{\infty} m e^{-im(\sigma-\sigma')}$$

We have also used Eq. (2.50) to relate ℓ_s and the string tension T. This gives us the commutation relation for the modes

$$\left[\alpha_m^\mu, \alpha_n^\nu \right] = \eta^{\mu\nu} m \delta_{m+n,0}$$

Equation (4.5) can be used to show that the α_m^μ and $\bar{\alpha}_n^\nu$ commute. We can write all of the commutation relations for the modes of the closed string as

$$\left[\alpha_m^\mu, \alpha_n^\nu \right] = m\eta^{\mu\nu}\delta_{m+n,0} \qquad \left[\bar{\alpha}_m^\mu, \bar{\alpha}_n^\nu \right] = m\eta^{\mu\nu}\delta_{m+n,0} \qquad \left[\alpha_m^\mu, \bar{\alpha}_n^\nu \right] = 0 \quad (4.9)$$

In the chapter quiz you will also derive a commutation relation for the center-of-mass position and momentum of the string

$$[x^\mu, p^\nu] = i\eta^{\mu\nu}$$

COMMUTATION RELATIONS FOR THE OPEN STRING

In the case of the open string, it can be shown that together with $[x^\mu, p^\nu] = i\eta^{\mu\nu}$, the commutation relations are

$$\left[\alpha_m^\mu, \alpha_n^\nu \right] = m\eta^{\mu\nu}\delta_{m+n,0} \tag{4.10}$$

THE OPEN STRING SPECTRUM

With the commutation relations in hand, we can proceed to find the states of the string. Because the open string case is simpler, we consider this first. Notice that in our quantization procedure where we have imposed commutation relations on the modes α_m^μ (and $\bar{\alpha}_n^\nu$ for the closed string), what we have done is promote them to *operators*. By extension, since the $X^\mu(\sigma, \tau)$ are defined in terms of α_m^μ, the $X^\mu(\sigma, \tau)$ are now to be thought of as operators as well. Therefore the next task in our program is to determine the state space of the system, that is the states upon which the α_m^μ and by extension the $X^\mu(\sigma, \tau)$ act. This procedure is really pretty similar to what you're used to from your previous studies of quantum theory.

The first item to notice is that the commutation relations have some similarity to the harmonic oscillator you learned about in first semester quantum mechanics. Temporarily dispensing with our convention of setting $\hbar = 1$, recall that we can define the creation and annihilation operators for the harmonic oscillator as

$$\hat{a}^\dagger = \sqrt{\frac{m\omega}{2\hbar}}\left(\hat{x} - \frac{i}{m\omega}\hat{p}\right) \qquad \hat{a} = \sqrt{\frac{m\omega}{2\hbar}}\left(\hat{x} + \frac{i}{m\omega}\hat{p}\right)$$

These operators satisfy the commutation relation:

$$[\hat{a}, \hat{a}^\dagger] = 1 \tag{4.11}$$

The hamiltonian of the system is given by

$$\hat{H} = \hbar\omega\left(\hat{a}^\dagger\hat{a} + \frac{1}{2}\right)$$

We introduce the *number operator* $\hat{N} = \hat{a}^\dagger\hat{a}$ which has eigenstates $|n\rangle$:

$$\hat{N}|n\rangle = n|n\rangle \tag{4.12}$$

where $n = 0, 1, 2, \ldots$. A system consisting of an infinite collection of harmonic oscillators is called a *fock space*.

Continuing, the number operator and its eigenstates allow us to write down the quantized energy levels of the harmonic oscillator, which are given by

$$\hat{H}|n\rangle = \hbar\omega\left(\hat{N} + \frac{1}{2}\right)|n\rangle = \hbar\omega\left(n + \frac{1}{2}\right)|n\rangle = E_n|n\rangle$$

You should also recall that the system has a ground state, which is the lowest possible energy state. This is denoted by $|0\rangle$.

A comparison of Eqs. (4.9) and (4.11) indicates that we have a similar system in the case of the string. This should not be surprising, since what else would you expect for a vibrating string, except a system that resembles a harmonic oscillator. Let's continue forward. Since $[\alpha_m^\mu, \alpha_n^\nu] = \eta^{\mu\nu} m \delta_{m+n,0}$, we can write

$$\left[\alpha_m^\mu, \alpha_n^\nu\right] = \left[\alpha_m^\mu, \alpha_{-m}^\nu\right] = m\eta^{\mu\nu} \qquad (4.13)$$

At this point, it is important to take a step back and recognize a key fact. The commutation relation not only bears some resemblance to the harmonic oscillator of quantum mechanics, but there is a very important difference. Notice that the presence of the metric $\eta_{\mu\nu}$ means that we can have negative commutators. This is the case for the time components. That is, since $\eta_{00} = -1$, it follows that

$$\left[\alpha_m^0, \alpha_{-m}^0\right] = -m$$

This is going to turn out to be important because it can lead to negative norm states.

Now, in analogy with the harmonic oscillator from ordinary quantum mechanics, we define a number operator. These are given in terms of the modes as

$$N_m = \alpha_{-m} \cdot \alpha_m$$

where we take $m \geq 1$. The eigenstates of the number operator satisfy

$$N_m \left| i_m \right\rangle = i_m \left| i_m \right\rangle$$

The *total number operator* is defined by summing over all possible $N_m = \alpha_{-m} \cdot \alpha_m$:

$$N = \sum_{m=1}^{\infty} N_m = \sum_{m=1}^{\infty} \alpha_{-m} \cdot \alpha_m \qquad (4.14)$$

Following a procedure used in elementary quantum mechanics, we can use $N_m = \alpha_{-m} \cdot \alpha_m$ together with $[\alpha_m^\mu, \alpha_n^\nu] = [\alpha_m^\mu, \alpha_{-m}^\nu] = m\eta_{\mu\nu}$ to show that the α_{-m}^μ and α_m^μ are raising and lowering operators, respectively. This follows from

$$N_m \left(\alpha_m \left| i_m \right\rangle \right) = \alpha_{-m} \cdot \alpha_m \left(\alpha_m \left| i_m \right\rangle \right)$$
$$= (\alpha_m \cdot \alpha_{-m} - m)\left(\alpha_m \left| i_m \right\rangle \right)$$
$$= \alpha_m (\alpha_{-m}\alpha_m)\left| i_m \right\rangle - m\alpha_m \left| i_m \right\rangle$$
$$= \alpha_m i_m \left| i_m \right\rangle - m\alpha_m \left| i_m \right\rangle = (i_m - m)\left(\alpha_m \left| i_m \right\rangle \right)$$

Hence, α_m^μ acts like a lowering operator. A similar exercise shows that the negative frequency mode α_{-m}^μ, $m \geq 1$ acts like a raising operator:

$$N_m \left(\alpha_{-m} | i_m \rangle \right) = (i_m + m) \left(\alpha_{-m} | i_m \rangle \right)$$

The lowering operator α_m^μ destroys the vacuum or ground state, which is the state with $i_m = 0$:

$$\alpha_m^\mu | 0 \rangle = 0 \tag{4.15}$$

We can construct higher-energy states using the raising operators which are the negative modes, α_{-m}^μ, $m \geq 1$, that is, $\alpha_{-1}^\mu | 0 \rangle$, $\alpha_{-1}^\mu \alpha_{-1}^\mu | 0 \rangle$, $\alpha_{-1}^\mu \alpha_{-2}^\mu | 0 \rangle$, and so on. A string state also carries momentum, so we can label a state by $| i_m, k \rangle$. Considering the ground state, supposing that the string carries momentum k^μ, the momentum operator acts as

$$p^\mu | 0, k \rangle = k^\mu | 0, k \rangle \tag{4.16}$$

Earlier we remarked that since the Minkowski metric $\eta^{\mu\nu}$ appears in the commutation relations, negative norm states can exist. We can demonstrate this explicitly as follows. Consider the first excited state with momentum k^μ, that is, $\alpha_{-1}^0 | 0, k \rangle$. Using $(\alpha_{-1}^0)^\dagger = \alpha_1^0$ we find the norm of this state to be

$$\left| \alpha_{-1}^0 | 0, k \rangle \right| = \langle 0, k | \alpha_1^0 \alpha_{-1}^0 | 0, k \rangle = -1 \tag{4.17}$$

We can rid the theory of the negative norm states by applying the Virasoro constraints. The classical expressions for the Virasoro constraints are

$$L_m = \frac{1}{2} \sum_n \alpha_{m-n} \cdot \alpha_n \qquad \bar{L}_m = \frac{1}{2} \sum_n \bar{\alpha}_{m-n} \cdot \bar{\alpha}_n \tag{4.18}$$

In the quantum theory, the Virasoro constraints are promoted to Virasoro *operators*. However, since the modes must satisfy the given commutation relations, some care must be applied when writing the Virasoro operators as derived from the classical expressions. The technique of *normal ordering* is used. This will ensure that the eigenvalues of the Virasoro operators will be finite. The prescription of normal ordering is simple:

- Move all lowering operators (positive frequency modes) to the right.
- Move all raising operators (negative frequency modes) to the left.

A normal ordered product is denoted using two colons, that is $: a^\dagger a :$. In the case of the Virasoro operator, we write

$$L_m = \frac{1}{2} \sum_n : \alpha_{m-n} \cdot \alpha_n :$$

L_m will lower the eigenvalue of the number operator by m. Looking at the commutation relation Eq. (4.10), you can see that α^μ_{m-n} and α^μ_n commute when $m \neq 0$. This means that when $m \neq 0$ we can simply move raising and lowering operators where we want in the expression for the Virasoro operator because no extra terms will be added from the commutator. Normal ordering L_0 gives

$$L_0 = \frac{1}{2} \alpha_0^2 + \sum_{n=1}^{\infty} \alpha_{-n} \cdot \alpha_n \tag{4.19}$$

Now, to get this result, note that

$$\frac{1}{2} \sum_{n=-\infty}^{\infty} \alpha_{-n} \cdot \alpha_n = \frac{1}{2} \alpha_0 \cdot \alpha_0 + \frac{1}{2} \sum_{n=1}^{\infty} \alpha_{-n} \cdot \alpha_n + \frac{1}{2} \sum_{n=1}^{\infty} \alpha_n \cdot \alpha_{-n}$$

$$= \frac{1}{2} \alpha_0 \cdot \alpha_0 + \sum_{n=1}^{\infty} \alpha_{-n} \cdot \alpha_n + \frac{1}{2} \sum_{\mu=0}^{D-1} \eta^\mu{}_\mu \sum_{n=1}^{\infty} n = \frac{1}{2} \alpha_0 \cdot \alpha_0 + \sum_{n=1}^{\infty} \alpha_{-n} \cdot \alpha_n + \frac{D}{2} \sum_{n=1}^{\infty} n$$

To get from the first to the second line, the commutator for the modes was used together with the fact that $\alpha_{-n} \cdot \alpha_n$ represents the dot product in D space-time dimensions. To get the normally ordered result we have thrown out the sum $D/2 \sum_{n=1}^{\infty} n$. At first glance, you might think that this sum is infinite, but regularization can be used to compute its finite value. To see how this works, recall the geometric series:

$$\sum_n a r^n = \frac{a}{1-r}$$

Taking the derivative, we have

$$\frac{d}{dr} \sum_n a r^n = \sum_n a n r^n = \frac{d}{dr} \frac{a}{1-r} = -\frac{a}{(1-r)^2}$$

Now let $a = 2\pi\varepsilon / \ell$ and multiply $\sum_{n=1}^{\infty} n$ by e^{-a}, then

$$\frac{d}{da} \sum_n (e^{-a})^n = \sum_n n (e^{-a})^n = \frac{d}{da} \frac{1}{1-e^{-a}} = \frac{e^{-a}}{(1-e^{-a})^2} = \frac{1}{a^2} - \frac{1}{12}$$

It follows that $D/2 \sum_{n=1}^{\infty} n = D/2(-1/12) = -D/24$. Another undetermined constant piece is missing from the difference between the general expression of L_0 and the normal ordered expression of L_0. This is a normal ordering constant which is denoted by a. Therefore in any calculation L_0 is replaced by $L_0 - a$, where a is a constant.

The point of all this is to write down the commutation relations for the Virasoro operators. Using Eq. (4.10), one finds that

$$[L_m, L_n] = (m-n)L_{m+n} + \frac{D}{12}(m^3 - m)\delta_{m+n, 0} \tag{4.20}$$

We call this commutation relation the *Virasoro algebra with central extension.* The *central charge* is the space-time dimension D which has shown up in the second term on the right-hand side. This is also the number of free scalar fields on the worldsheet. It is clear that if $m = 0, \pm 1$ the central extension term will vanish. This singles out L_1, L_0, and L_{-1} which form a closed subalgebra. We call this the SL (2, R) algebra.

The Virasoro operators can be used to eliminate unphysical states (i.e., negative norm states) from the theory by requiring that the expectation value of $L_0 - a$ vanishes for a physical state $|\psi\rangle$. That is, we impose the constraint

$$\langle \psi | L_m - a\delta_{m,0} | \psi \rangle = 0$$

for $m \geq 0$. The term $a\delta_{m,0}$ takes care of the fact that we only need the normal ordering constant a in the case of L_0. To eliminate negative norm states, specific conditions must be put on a and D, which is the origin of the "extra dimensions" in string theory. In particular, it can be shown that negative norm states can be eliminated if

$$a = 1 \qquad D = 26 \tag{4.21}$$

The reason that $a = 1$ is chosen is a bit beyond the scope we want to cover in this book, see the references if interested in the proof.

We can proceed further to obtain a mass operator. First recall that Einstein's equation tells us

$$p^\mu p_\mu + m^2 = 0 \qquad \Rightarrow m^2 = -p^\mu p_\mu$$

To obtain a formula for the mass operator in string theory, which is denoted by M^2, we use the condition on physical states. Taking $L_0 = 1/2\alpha_0^2 + \sum\limits_{n=1}^{\infty} \alpha_{-n} \cdot \alpha_n$, using $L_0 - a$ with $a = 1$, we arrive at the condition

$$(L_0 - a)|\psi\rangle = 0 \Rightarrow \left(\frac{1}{2}\alpha_0^2 + \sum_{n=1}^{\infty} \alpha_{-n} \cdot \alpha_n - 1 \right)|\psi\rangle = 0$$

$$\Rightarrow \frac{1}{2}\alpha_0^2 + \sum_{n=1}^{\infty} \alpha_{-n} \cdot \alpha_n - 1 = 0$$

The first term in this expression is nothing other than the mass squared: $(1/2)\alpha_0^2 = -\alpha' M^2$ where $\alpha' = 1/(2\pi T)$. So, in bosonic string theory the "mass shell" condition becomes

$$M^2 = \frac{1}{\alpha'}(N - 1) \tag{4.22}$$

where N is the total number operator. The term $\sqrt{2\pi T}$ sets the energy scale of the theory, it is taken to be on the order of the Planck mass. This is the origin of the high energy scale of string theory.

It can be shown that

$$M^2 = \frac{1}{\alpha'}\left(N - \frac{D-2}{24} \right)$$

Notice that setting $a = 1$ forces us to take $D = 26$. The number operator acts on the ground state as

$$N|0\rangle = 0$$

Hence the mass of the ground state is

$$M^2|0\rangle = \frac{1}{\alpha'}\left(N - \frac{D-2}{24} \right)|0\rangle = -\frac{1}{\alpha'}\frac{D-2}{24}|0\rangle = -\frac{1}{\alpha'}|0\rangle$$

So the ground state of bosonic string theory in the open string case has *negative mass*. This means that the ground state is a Tachyon. This is an unphysical state which travels faster than the speed of light. Consistency of bosonic string theory requires that we choose $a = 1$, so the Tachyon cannot be removed from the theory.

It will turn out that the introduction of supersymmetry (that is the introduction of fermionic states) into the theory will get rid of the Tachyon, giving us a realistic string theory. We will see that this also changes the number of space-time dimensions.

Now let's consider the mass of the first excited state. The first excited state is $|i\rangle = \alpha_{-1}^i|0\rangle$ where i is a spatial index. Here, $i = 1, ..., D-2$, and the state transforms as a vector in space-time. You will recall from your studies of quantum field theory that a vector is a spin-1 particle that in general has $D-1$ components, the fact that this state has $D-2$ components implies that it is a massless state. An example of a massless vector is the photon, it only has transverse components of spin. This explains why there are $D-2$ components rather than $D-1$. The mass of the state is

$$M^2\alpha_{-1}^i|0\rangle = \frac{1}{\alpha'}\left(1 - \frac{D-2}{24}\right)\alpha_{-1}^i|0\rangle = \frac{1}{\alpha'}\left(\frac{26-D}{24}\right)\alpha_{-1}^i|0\rangle$$

In order for the state to be massless, $26 - D/24$ must vanish, once again setting the number of space-time dimensions D to 26. Physicists refer to the bosonic string theory with $a = 1$, $D = 26$ as *critical* and call $D = 26$ the *critical dimension*.

CLOSED STRING SPECTRUM

In the case of the closed string, things are a little more complicated than what you're used to from the harmonic oscillator in ordinary quantum theory due to the fact that we have a second commutation relation in addition to $\left[\alpha_m^\mu, \alpha_n^\nu\right] = m\,\eta_{\mu\nu}\delta_{m+n,0}$ that must be satisfied, namely, $\left[\bar\alpha_m^\mu, \bar\alpha_n^\nu\right] = m\,\eta_{\mu\nu}\delta_{m+n,0}$. What this is going to mean is that we need to define *two* number operators. These are defined by infinite sums over the modes:

$$N_R = \sum_{m=1}^{\infty}\bar\alpha_{-m}\bar\alpha_m \qquad N_L = \sum_{m=1}^{\infty}\alpha_{-m}\alpha_m \qquad (4.23)$$

Together with the momentum operator p^μ, the number operators N_R and N_L serve to characterize the state of a closed string. Let us denote a state by $|n, k\rangle$. As in the open string case, the momentum operator will act according to

$$p^\mu|0, k\rangle = k^\mu|0, k\rangle \qquad (4.24)$$

Therefore, the state $|0, k\rangle$ of the string carries momentum k^μ. Turning our attention to the number operators, first let's specify the action of the raising and lowering (creation and annihilation) operators. This follows what we did in the open string case as well. Keeping $m \geq 1$, we define these as follows:

- α_m^μ is a lowering operator.
- α_{-m}^μ is a raising operator.

Similar roles are played by $\bar{\alpha}_m^\mu$ and $\bar{\alpha}_{-m}^\mu$. We define the ground state by $|0, 0\rangle$ which is often abbreviated by $|0\rangle$. Then, the lowering operators satisfy a relation familiar from ordinary quantum mechanics:

$$\alpha_m^\mu |0\rangle = 0 \qquad \bar{\alpha}_m^\mu |0\rangle = 0 \qquad (4.25)$$

Note also that we can define the ground state with momentum k^μ by writing $|k\rangle$. The raising operators α_{-m}^μ and $\bar{\alpha}_{-n}^\mu$ act to raise the eigenvalues of the number operators N_L and N_R by m and n, respectively. So if $i(\mu, m)$ and $\bar{i}(\mu, m)$ are integers:

$$\left| i(\mu, m)\bar{i}(\mu, m), k \right\rangle = \prod_{m \geq 1} \prod_{\mu=0}^{D-1} \left(\alpha_{-m}^\mu \right)^i \left(\bar{\alpha}_{-m}^\mu \right)^{\bar{i}} |0, k\rangle$$

The number operators N_L and N_R act on this state as follows:

$$N_L \left| i(\mu, m)\bar{i}(\mu, m), k \right\rangle = \sum_{\mu,m} m\, i(\mu, m) \left| i(\mu, m)\bar{i}(\mu, m), k \right\rangle$$

$$N_R \left| i(\mu, m)\bar{i}(\mu, m), k \right\rangle = \sum_{\mu,m} m\, \bar{i}(\mu, m) \left| i(\mu, m)\bar{i}(\mu, m), k \right\rangle$$

Once again, we will have negative norm states due to the presence of the metric in the commutation relations. This situation can be dealt with in the same manner used in the open string case. So we won't go into great detail and simply state the results, noting that in the following we take $m \geq 0$. We proceed by introducing normal ordered Virasoro operators, but this time must include $\bar{L}_m$ as well:

$$L_m = \frac{1}{2} \sum_n : \alpha_{m-n} \cdot \alpha_n : \qquad \bar{L}_m = \frac{1}{2} \sum_n : \bar{\alpha}_{m-n} \cdot \bar{\alpha}_n : \qquad (4.26)$$

These operators satisfy commutation relations called the Virasoro algebra:

$$[L_m, L_n] = (m-n)L_{m+n} + \frac{D}{12}(m^3 - m)\delta_{m+n,0}$$

$$[\bar{L}_m, \bar{L}_n] = (m-n)\bar{L}_{m+n} + \frac{D}{12}(m^3 - m)\delta_{m+n,0}$$

$$(4.27)$$

If the following relations are satisfied by a state $|\psi\rangle$:

$$(L_m - a\delta_{m,0})|\psi\rangle = 0 \qquad (\bar{L}_m - a\delta_{m,0})|\psi\rangle = 0 \qquad (4.28)$$

then the state $|\psi\rangle$ is a physical state. That is, if $m > 0$, then the Virasoro operator annihilates the physical state $L_m|\psi\rangle = 0$, $\bar{L}_m|\psi\rangle = 0$. The condition satisfied when $m = 0$ is the "mass shell" condition, $(L_0 - a)|\psi\rangle = 0$, $(\bar{L}_0 - a)|\psi\rangle = 0$. Once again, it can be shown that negative norm states can be avoided in the theory provided that $a = 1$, $D = 26$. The Virasoro operators L_0 and $\bar{L}_0$ can be written in terms of the number operators as follows:

$$L_0 = \frac{1}{8\pi T} p^\mu p_\mu + N_L \qquad \bar{L}_0 = \frac{1}{8\pi T} p^\mu p_\mu + N_R$$

The sum and difference of L_0 and $\bar{L}_0$ annihilate the physical states:

$$(L_0 + \bar{L}_0 - 2a)|\psi\rangle = 0 \qquad (L_0 - \bar{L}_0)|\psi\rangle = 0$$

The constraint $(L_0 - \bar{L}_0)|\psi\rangle = 0$ is called *level matching*. Using the Einstein relations, we can arrive at an expression for a mass operator:

$$M^2 = -p^\mu p_\mu = \frac{2}{\alpha'}(N_L + N_R - 2) \qquad (4.29)$$

The ground state of the closed bosonic string is $|0, k\rangle$, found when $N_L = N_R = 0$. This is a tachyon that satisfies

$$M^2 = -\frac{4}{\alpha'}$$

The next case to consider is $N_L = N_R = 1$, which is the first excited state. Here we get hints that string theory is a unified theory. The first excited states are massless, so $M^2 = 0$. They are derived from the ground state in the following way:

$$\varepsilon_{\mu\nu}(k)\alpha_{-1}^\mu \bar{\alpha}_{-1}^\nu |0, k\rangle$$

The object $\varepsilon_{\mu\nu}(k)$ is a *tensor* which can be decomposed into symmetric $\varepsilon_{(\mu\nu)}(k)$ and antisymmetric $\varepsilon_{(\mu\nu)}(k)$ parts (see *Relativity Demystified* if you aren't sure about this). The symmetric part corresponds to a massless spin-2 particle which is the *graviton*. The linearized metric $g_{\rho\sigma} = \eta_{\rho\sigma} + \varepsilon_{\rho\sigma}(x)$ satisfies the linearized Einstein equations $\partial_\mu \partial^\mu \varepsilon_{\rho\sigma}(x) = 0$, and $\partial^\mu \varepsilon_{\mu\nu}(x) = 0$. By taking the Fourier transform of $\varepsilon_{(\mu\nu)}(k)$, it can be shown that these equations are satisfied.

The trace $\varepsilon^\mu_{\ \mu}(k)$, which defines a scalar, is also important. This corresponds to a massless scalar particle called the *dilaton*.

Light-Cone Quantization

We now turn our attention to a different method of quantization. We began the chapter with a discussion of covariant quantization. This method is a straightforward application of the imposition of commutation relations. We took this approach first because it may seem familiar from your studies of ordinary quantum mechanics. In addition, it preserves the Lorentz invariance of the theory. Physicists say that this approach is "manifestly" Lorentz invariant, colloquially meaning that the Lorentz invariance is obvious. The technique has the disadvantage in that negative norm states appear. Although this is a problem, it is instructive to go through the process of eliminating the negative norm states.

Another approach is possible which avoids the negative norm states at the cost of losing manifest Lorentz invariance. This is called *light-cone quantization*. We briefly discuss it here, considering the open string case.

We begin by using light-cone coordinates, which were introduced in Chap. 2 in Eq. (2.34):

$$X^\pm = \frac{X^0 \pm X^{D-1}}{\sqrt{2}} \tag{4.30}$$

The remaining coordinates X^i are transverse coordinates. The center of mass coordinate x^μ and momentum p^μ are also written as light-cone coordinates. In the light-cone gauge, we choose

$$X^+ = x^+ + \ell_s^2 p^+ \tau \tag{4.31}$$

which leads to $\alpha_n^+ = 0$ for $n \neq 0$, that is, the modes are zero for X^+. The Virasoro constraints will lead us to a description based on transverse oscillators. We have the freedom to set $\ell_s^2 p^+ = 1$, giving the center of mass position as

$$\bar{X}^\mu = x^\mu + \frac{p^\mu}{p^+}\tau$$

The Virasoro constraint becomes

$$\dot{X}^- \pm X^{-\prime} = \frac{1}{2}(\dot{X}^i \pm X^{i\prime})^2 \tag{4.32}$$

So we have a relation between light-cone coordinates and the transverse coordinates. The mode expansion of the worldsheet coordinates for an open string is given by

$$X^\mu = x^\mu + \ell_s^2 p^\mu \tau + i\ell_s \sum_{n\neq 0} \frac{\alpha_n^\mu}{n} e^{-in\tau} \cos n\sigma$$

In particular

$$X^- = x^- + \ell_s^2 p^- \tau + i\ell_s \sum_{n\neq 0} \frac{\alpha_n^-}{n} e^{-in\tau} \cos n\sigma$$

We can then solve for the nonzero modes of X^- in terms of the transverse oscillators. These are

$$\alpha_n^- = \frac{1}{p^+ \ell_s}\left(\frac{1}{2}\sum_{i=1}^{D-2}\sum_{m=-\infty}^{\infty} : \alpha_{n-m}^i \alpha_m^i : -a\delta_{n,0} \right) \tag{4.33}$$

In the case of the zeroth mode, we can derive an expression for the hamiltonian

$$H = p^- = \frac{1}{2p^+}\left(p^i p^i + \frac{1}{\alpha'}\sum_{n\neq 0}\alpha_{-n}^i \alpha_n^i \right) \tag{4.34}$$

Define a conjugate momentum P^μ. The system is quantized by imposing commutation relations on the transverse components of position and momentum:

$$[x^i, p^j] = i\delta^{ij}, \quad [X^i(\sigma), P^j(\sigma')] = i\delta^{ij}\delta(\sigma - \sigma') \tag{4.35}$$

The mass-shell condition becomes

$$M^2 = 2p^+ p^- - \sum_{i=1}^{D-2} p_i^2 = \frac{2}{\ell_s^2}(N - a) \tag{4.36}$$

The number operator is given by

$$N = \sum_{i=1}^{D-2}\sum_{n=1}^{\infty}\alpha_{-n}^i \alpha_n^i \tag{4.37}$$

Normal ordering leads to

$$\frac{1}{2}\sum_{i=1}^{D-2}\sum_{n=-\infty}^{\infty}\alpha_{-n}^{i}\alpha_{n}^{i} = \frac{1}{2}\sum_{i=1}^{D-2}\sum_{n=-\infty}^{\infty}:\alpha_{-n}^{i}\alpha_{n}^{i}:+\frac{D-2}{2}\sum_{n=1}^{\infty}n$$

The regularization trick can be applied to make the second term finite. We find

$$\frac{D-2}{2}\sum_{n=1}^{\infty}n = -\frac{D-2}{24}$$

Again taking $a = 1$, one finds $D = 26$.

Summary

In the previous two chapters, we constructed a relativistic theory of the string, the classical theory. In this chapter we have introduced the simplest possible quantum extension of the classical theory. This is a theory that consists only of bosons. While the theory is not realistic since it does not include fermionic states, it is easier to deal with and introduces important concepts and methods that will play a role in the full quantum theory. The classical theory was quantized using two different methods. The first method, called covariant quantization, is a straightforward approach that imposes commutation relations on the X^{μ} and their conjugate momenta. This leads to negative norm states. The Virasoro constraints are imposed to rid the theory of these states. When this is done, we find that the theory must have 26 space-time dimensions. We concluded the chapter with a different approach, known as light-cone quantization.

Quiz

1. Explicitly calculate the commutators $[\partial_{+}X^{\mu}(\sigma, \tau), \partial'_{-}X^{\nu}(\sigma', \tau)]$ and $[\partial_{-}X^{\mu}(\sigma, \tau), \partial'_{+}X^{\nu}(\sigma', \tau)]$.

2. Consider the closed string and explicitly calculate $[x^{\mu}, p^{\nu}]$.

3. Consider the first excited state of the closed string $\varepsilon_{\mu\nu}(k)\alpha_{-1}^{\mu}\bar{\alpha}_{-1}^{\nu}|0, k\rangle$. Using the condition satisfied by physical states $|\psi\rangle$, in particular $L_{1}|\psi\rangle = \bar{L}_{1}|\psi\rangle = 0$, find $\varepsilon_{\mu\nu}k^{\mu}$.

4. Let $\alpha_{-1}^{i}|0, k\rangle$ be the state of an open string and suppose that the normal ordering constant a is undetermined. What is the mass of this state?

5. In the light-cone gauge, use $[x^{i}, p^{j}] = i\delta^{ij}$, $[X^{i}(\sigma), P^{j}(\sigma')] = i\delta^{ij}\delta(\sigma - \sigma')$ to find a commutation relation for the transverse modes, $[\alpha_{m}^{i}, \alpha_{n}^{j}]$.

6. Consider light-cone quantization for the closed string case. What additional commutation relation do you think should be imposed for the modes?

CHAPTER 5

Conformal Field Theory Part I

In this chapter we study *conformal field theory,* an area of quantum field theory that relies heavily on complex variables. In this chapter, we will introduce some of the basic concepts of conformal field theory. The topic will be expanded and utilized in many areas in the rest of the book. In the next chapter, we discuss other aspects of conformal field theory along with BRST quantization.

Conformal field theory is an important tool used in the analysis of perturbative string theory, so it plays a central role in our task at hand (understanding the physics of quantized strings). In particular, since the worldsheet can be described using two coordinates (τ, σ) two-dimensional conformal field theory is used. The theory of complex variables plays an important role in the study of theoretical physics, and string theory is no exception. If you are not familiar with complex variables you should take time out to study the topic before proceeding any further. You can do so using my book *Complex Variables Demystified,* also published by McGraw-Hill.

In the theory of complex variables, a *conformal transformation* is one that maps a region of the complex plane to a new more convenient region while preserving angles, but not lengths. For example, you can map the unit disk to the upper-half plane using a conformal transformation.

The notion that angles are preserved is a geometric interpretation and leads us to the notion of a conformal transformation of space-time coordinates. Let us consider a transformation of the space-time coordinates such that $x \rightarrow x'$. In general, the metric $g_{\mu\nu}(x)$ is found to transform in the following way:

$$g_{\mu\nu}{}'(x') = \frac{\partial x^\alpha}{\partial x'^\mu} \frac{\partial x^\beta}{\partial x'^\nu} g_{\alpha\beta}(x) \tag{5.1}$$

Now consider a function of the space-time coordinates given by $\Omega(x)$. If the metric transforms in the following way:

$$g_{\mu\nu}{}'(x') = \Omega(x) g_{\mu\nu}(x) \tag{5.2}$$

Then we have a conformal transformation of the metric. Notice that $\Omega(x)$ acts as a scaling factor, hence it will preserve angles but not lengths. If a metric is related to the flat Minkowski metric as $g_{\mu\nu} = \Omega(x)\eta_{\mu\nu}(x)$ we say that the metric is *conformally flat*.

To see how a conformal transformation preserves angles, consider two tangent vectors **u** and **v**. Using the metric, the angle between them is given by

$$\cos\theta = \frac{g(\mathbf{u}, \mathbf{v})}{\sqrt{g(\mathbf{u}, \mathbf{u})\, g(\mathbf{v}, \mathbf{v})}}$$

Now we apply the transformation given by Eq. (5.2) and find

$$\cos\theta \rightarrow \frac{g'(\mathbf{u}, \mathbf{v})}{\sqrt{g'(\mathbf{u}, \mathbf{u})\, g'(\mathbf{v}, \mathbf{v})}} = \frac{\Omega(x)g(\mathbf{u}, \mathbf{v})}{\sqrt{\Omega(x)g(\mathbf{u}, \mathbf{u})\Omega(x)g(\mathbf{v}, \mathbf{v})}} = \frac{g(\mathbf{u}, \mathbf{v})}{\sqrt{g(\mathbf{u}, \mathbf{u})\, g(\mathbf{v}, \mathbf{v})}}$$

Hence a conformal transformation preserves angles.

A conformal field theory is a quantum field theory that is invariant under conformal transformations. These theories are *Euclidean quantum field theories,* meaning that a Euclidean metric is used. The symmetry group of such a theory will contain Euclidean symmetries (we will review those in a moment) along with local conformal transformations. It turns out that two-dimensional conformal field theories are of particular use. Two-dimensional conformal field theories have an infinite number of conserved charges.

There are two important properties of conformal field theories. These can be summed up by saying that a conformal field theory is *scale invariant*. This manifests itself in two ways:

- Conformal field theories have no length scale.
- Conformal field theories have no mass scale.

To see why this is important, we can consider the quantum field theory of a scalar field. Let $\phi(x)$ be a scalar field in d space-time dimensions. Consider a rescaling of the coordinates which we write as a *scale transformation*:

$$x' = \lambda x \tag{5.3}$$

Under a scale transformation, a scalar field $\phi(x)$ transforms using the *classical scaling dimension* $\Delta = (d-2)/2$ as follows:

$$\phi(x) \to \phi'(\lambda x) = \lambda^{-\Delta}\phi(x) = \lambda^{-\frac{d}{2}+1}\phi(x)$$

For a field theory to be scale invariant, we require that the action be invariant under this transformation. This is, in fact, true when we consider a free, massless scalar field. The action in this case is

$$S = \int d^d x\, \partial_\mu \phi \partial^\mu \phi$$

Under the transformation $x' = \lambda x$, it is clear that $dx' = \lambda dx$, that is,

$$d^d x' = d(\lambda x_0)d(\lambda x_1)...d(\lambda x_{d-1}) = (\lambda \cdot \lambda ... \lambda)d(x_0)d(x_1)...d(x_{d-1}) = \lambda^d d^d x$$

Now recall that ∂_μ is shorthand for $\partial/(\partial x^\mu)$. So we pick up a copy of λ under a scale transformation:

$$\frac{\partial}{\partial x^\mu} \to \frac{\partial}{\partial(\lambda x^\mu)} = \frac{1}{\lambda}\frac{\partial}{\partial x^\mu}$$

So we have

$$\partial_\mu \phi \partial^\mu \phi \to \left(\frac{1}{\lambda}\right)\partial_\mu(\phi')\left(\frac{1}{\lambda}\right)\partial^\mu(\phi') = \left(\frac{1}{\lambda^2}\right)\partial_\mu\left(\lambda^{-\frac{d}{2}+1}\phi\right)\partial^\mu\left(\lambda^{-\frac{d}{2}+1}\phi\right)$$

$$= \left(\frac{1}{\lambda^2}\right)\lambda^{-d+2}\partial_\mu \phi \partial^\mu \phi = \lambda^{-d}\partial_\mu \phi \partial^\mu \phi$$

where we used $\phi(x) \rightarrow \phi'(\lambda x) = \lambda^{-(d/2)+1}\phi(x)$. It follows that the action is unchanged under a scale transformation:

$$S' = \int d^d x' \partial_{\mu'} \phi' \partial^{\mu'} \phi' = \int \lambda^d d^d x \, \lambda^{-d} \partial_\mu \phi \partial^\mu \phi = \int d^d x \, \partial_\mu \phi \partial^\mu \phi = S$$

The problem is that a scale transformation does not change a fixed quantity like mass. Let's consider a free scalar field with mass m. The action is

$$S = \int d^d x \, (\partial_\mu \phi \partial^\mu \phi - m^2 \phi^2)$$

This action is *not* invariant under a scale transformation because

$$m^2 \phi'^2 = m^2 \lambda^{-d+2} \phi^2 \neq m^2 \phi^2$$

Often, quantum theory breaks scale invariance. A prototypical example is ϕ^4 theory for a massless field. The classical action is scale invariant:

$$S = \int d^4 x \left(\frac{1}{2} \partial_\mu \phi \partial^\mu \phi - \frac{g}{4!} \phi^4 \right)$$

The problem is, that renormalization introduces a fixed mass term to the theory. As a result scale invariance is broken. Conformal field theories provide a way out of this conflict by giving us quantum field theories that are scale *and* mass invariant.

The Role of Conformal Field Theory in String Theory

A question you should be asking yourself is why are conformal field theories important in string theory? It turns out that two-dimensional conformal field theories are very important in the study of worldsheet dynamics.

A string has internal degrees of freedom determined by its vibrational modes. The different vibrational modes of the string are interpreted as particles in the theory. That is, the different ways that the string vibrates against the background space-time determine what kind of particle the string is seen to exist as. So in one vibrational mode, the string is an electron, while in another, the string is a quark, for example. Yet a third vibrational mode is a photon.

As we have seen already, the vibrational modes of the string can be studied by examining the worldsheet, which is the two-dimensional surface. It turns out that when studying the worldsheet, the vibrational modes of the string are described by a conformal field theory.

If the string is closed, we have two vibrational modes (left movers and right movers) moving around the string independently. Each of these can be described by a conformal field theory. Since the modes have "direction" we call the theories that describe these two independent modes *chiral conformal field theories*. This will be important for open strings as well.

Wick Rotations

A *Euclidean metric* is simply a metric that resembles the distance measure from ordinary geometry. Let's try to clarify this point considering the simplified case of one-time dimension and one-space dimension. In special relativity, we distinguish between space and time with the use of a change of sign so that if we are using the signature $(-, +)$ $ds^2 = -dt^2 + dx^2$. So the two-dimensional Minkowski metric would be

$$\eta_{\mu\nu} = \begin{pmatrix} -1 & 0 \\ 0 & 1 \end{pmatrix}$$

What we're after with a Euclidean metric is describing things in a way that we could using ordinary geometry. In the *x-y* plane, the infinitesimal measure of distance is given by $dr^2 = dx^2 + dy^2$. This tells us that a Euclidean metric is one for which all quantities have the same sign. We can rewrite the Minkowski metric in this way by using what is known as a *Wick rotation*. Simply put, we make a transformation on the time coordinate by letting $t \rightarrow -it$. Then $dt \rightarrow -idt$ and it follows that $ds^2 = -(-idt)^2 + dx^2 = dt^2 + dx^2$, which is exactly what we want. In order to describe the worldsheet with coordinates (τ, σ) using a Euclidean metric, we make a Wick rotation $\tau \rightarrow -i\tau$.

In terms of the worldsheet coordinates (τ, σ), the metric is

$$ds^2 = -d\tau^2 + d\sigma^2$$

So we see that making a Wick rotation $\tau \rightarrow -i\tau$ changes this to

$$ds^2 = d\tau^2 + d\sigma^2$$

which is a Euclidean metric. Utilizing the Euclidean metric enables us to use conformal field theory on the string.

Complex Coordinates

A consequence of a Wick rotation is that the light-cone coordinates $(+, -)$ are replaced with complex coordinates $(z, \bar{z})$. The description of the worldsheet is transformed into complex variables by defining complex coordinates $(z, \bar{z})$ which are functions of the real variables (τ, σ). One way this can be done is as follows:

$$z = \tau + i\sigma \qquad \bar{z} = \tau - i\sigma \tag{5.4}$$

Let's use this definition to work out a few basic quantities and show how this simplifies analysis. Keep the Polyakov action in the back of your mind. Using the Euclidean metric the Polyakov action is written as

$$S_P = \frac{1}{4\pi\alpha'} \int d\tau d\sigma (\partial_\tau X^\mu \partial_\tau X_\mu + \partial_\sigma X^\mu \partial_\sigma X_\mu) \tag{5.5}$$

We're going to find out that going to complex variables will simplify the form of Eq. (5.5).

To transform coordinates we need to know how to compute derivatives with respect to the coordinates z and $\bar{z}$. This is easy enough. First we invert the coordinates Eq. (5.4):

$$\tau = \frac{z + \bar{z}}{2} \qquad \sigma = \frac{z - \bar{z}}{2i} \tag{5.6}$$

It follows that

$$\frac{\partial \tau}{\partial z} = \frac{\partial \tau}{\partial \bar{z}} = \frac{1}{2} \qquad \frac{\partial \sigma}{\partial z} = \frac{1}{2i} \qquad \frac{\partial \sigma}{\partial \bar{z}} = -\frac{1}{2i}$$

and so

$$\frac{\partial}{\partial z} = \frac{\partial \tau}{\partial z}\frac{\partial}{\partial \tau} + \frac{\partial \sigma}{\partial z}\frac{\partial}{\partial \sigma} = \frac{1}{2}\frac{\partial}{\partial \tau} + \frac{1}{2i}\frac{\partial}{\partial \sigma} = \frac{1}{2}\left(\frac{\partial}{\partial \tau} - i\frac{\partial}{\partial \sigma}\right) \tag{5.7}$$

The shorthand notation $\partial_z = \partial = 1/2(\partial_\tau - i\partial_\sigma)$ is usually used. It is also easy to see that

$$\frac{\partial}{\partial \bar{z}} = \partial_{\bar{z}} = \bar{\partial} = \frac{\partial \tau}{\partial \bar{z}}\frac{\partial}{\partial \tau} + \frac{\partial \sigma}{\partial \bar{z}}\frac{\partial}{\partial \sigma} = \frac{1}{2}\frac{\partial}{\partial \tau} - \frac{1}{2i}\frac{\partial}{\partial \sigma} = \frac{1}{2}\left(\frac{\partial}{\partial \tau} + i\frac{\partial}{\partial \sigma}\right) = \frac{1}{2}(\partial_\tau + i\partial_\sigma) \tag{5.8}$$

where we've introduced the abbreviation $\partial_{\bar{z}} = \bar{\partial}$. Now given that after a Wick rotation the metric for the (τ, σ) coordinate system is written as

$$g_{\alpha\beta} = \begin{pmatrix} 1 & 0 \\ 0 & 1 \end{pmatrix}$$

We can write down the metric in the new complex coordinates using Eq. (5.1). We have

$$g_{zz} = \partial_z \tau \partial_z \tau g_{\tau\tau} + \partial_z \tau \partial_z \sigma g_{\tau\sigma} + \partial_z \sigma \partial_z \tau g_{\sigma\tau} + \partial_z \sigma \partial_z \sigma g_{\sigma\sigma}$$

$$= \partial_z \tau \partial_z \tau + \partial_z \sigma \partial_z \sigma \tag{5.9}$$

$$= \left(\frac{1}{2}\right)\left(\frac{1}{2}\right) + \left(\frac{1}{2i}\right)\left(\frac{1}{2i}\right) = \frac{1}{4} - \frac{1}{4} = 0$$

Similarly, $g_{\bar{z}\bar{z}} = 0$. On the other hand

$$g_{z\bar{z}} = \partial_z \tau \partial_{\bar{z}} \tau g_{\tau\tau} + \partial_z \tau \partial_{\bar{z}} \sigma g_{\tau\sigma} + \partial_z \sigma \partial_{\bar{z}} \tau g_{\sigma\tau} + \partial_z \sigma \partial_{\bar{z}} \sigma g_{\sigma\sigma}$$

$$= \partial_z \tau \partial_{\bar{z}} \tau + \partial_z \sigma \partial_{\bar{z}} \sigma \tag{5.10}$$

$$= \left(\frac{1}{2}\right)\left(\frac{1}{2}\right) + \left(\frac{1}{2i}\right)\left(-\frac{1}{2i}\right) = \frac{1}{4} + \frac{1}{4} = \frac{1}{2} = g_{\bar{z}z}$$

In matrix form

$$g_{\mu\nu} = \begin{pmatrix} 0 & 1/2 \\ 1/2 & 0 \end{pmatrix} \tag{5.11}$$

The inverse metric, written with raised indices has components given by

$$g^{zz} = g^{\bar{z}\bar{z}} = 0 \qquad g^{z\bar{z}} = g^{\bar{z}z} = 2 \tag{5.12}$$

The corresponding matrix is

$$g^{\mu\nu} = \begin{pmatrix} 0 & 2 \\ 2 & 0 \end{pmatrix}$$

"Volume elements" in integrals can be written using coordinate transformations by including the determinant of the metric. Writing $d^2z = dz d\bar{z}$ and using $\sqrt{|\det g|}\, d^2z = d\tau d\sigma$ it follows that

$$d^2z = 2 d\tau d\sigma \tag{5.13}$$

Now consider the action

$$S = \frac{1}{2\pi\alpha'}\int d^2z\, \partial X^\mu\, \bar{\partial} X_\mu \tag{5.14}$$

This is, in fact, the Polyakov action [Eq. (5.5)] in a much simpler mathematical form. To see this, we can use Eq. (5.7) together with Eq. (5.13). Notice that

$$\partial X^\mu\, \bar{\partial} X_\mu = \left\{\frac{1}{2}(\partial_\tau - i\partial_\sigma)X^\mu\right\}\left\{\frac{1}{2}(\partial_\tau + i\partial_\sigma)X_\mu\right\}$$

$$= \frac{1}{4}(\partial_\tau X^\mu - i\partial_\sigma X^\mu)(\partial_\tau X_\mu + i\partial_\sigma X_\mu)$$

$$= \frac{1}{4}(\partial_\tau X^\mu \partial_\tau X_\mu + i\partial_\tau X^\mu \partial_\sigma X_\mu - i\partial_\sigma X^\mu \partial_\tau X_\mu + \partial_\sigma X^\mu \partial_\sigma X_\mu)$$

$$= \frac{1}{4}(\partial_\tau X^\mu \partial_\tau X_\mu + \partial_\sigma X^\mu \partial_\sigma X_\mu)$$

To move from the third to the fourth line, we used the fact we can raise and lower indices with the Euclidean metric. That is, $X_\mu = \delta_{\mu\nu}X^\nu = X^\mu$, so

$$-i\partial_\sigma X^\mu \partial_\tau X_\mu = -i\partial_\sigma X_\mu \partial_\tau X_\mu = -i\partial_\sigma X_\mu \partial_\tau X^\mu = -i\partial_\tau X^\mu \partial_\sigma X_\mu$$

and the middle terms cancel. Therefore

$$S = \frac{1}{2\pi\alpha'}\int d^2z\, \partial X^\mu\, \bar{\partial} X_\mu = \frac{1}{2\pi\alpha'}\int 2 d\tau d\sigma\, \partial X^\mu\, \bar{\partial} X_\mu$$

$$= \frac{1}{2\pi\alpha'}\int 2 d\tau d\sigma \frac{1}{4}(\partial_\tau X^\mu \partial_\tau X_\mu + \partial_\sigma X^\mu \partial_\sigma X_\mu)$$

$$= \frac{1}{4\pi\alpha'}\int d\tau d\sigma (\partial_\tau X^\mu \partial_\tau X_\mu + \partial_\sigma X^\mu \partial_\sigma X_\mu) = S_P$$

We've shown that Eq. (5.14) is an equivalent way to write the Polyakov action. But it's much simpler, and it's much simpler to derive the equations of motion using this form. We can do this by varying the action [Eq. (5.14)] with respect to the coordinate X_μ. This is done by letting $X_\mu \to X_\mu + \delta X_\mu$. Then

$$S \to \frac{1}{2\pi\alpha'}\int d^2z\, \partial X^\mu\, \bar{\partial}(X_\mu + \delta X_\mu) = \frac{1}{2\pi\alpha'}\int d^2z\, \partial X^\mu(\bar{\partial}X_\mu + \bar{\partial}\delta X_\mu)$$

$$= \frac{1}{2\pi\alpha'}\int d^2z\, \partial X^\mu\, \bar{\partial}X_\mu + \frac{1}{2\pi\alpha'}\int d^2z\, \partial X^\mu(\bar{\partial}\delta X_\mu) = S + \delta S$$

We can obtain the equations of classical motion by requiring that $\delta S = 0$. Integrating by parts and discarding the boundary term:

$$\delta S = \frac{1}{2\pi\alpha'}\int d^2z\, \partial X^\mu(\bar{\partial}\delta X_\mu)$$

$$= -\frac{1}{2\pi\alpha'}\int d^2z\, \partial\bar{\partial}X^\mu(\delta X_\mu)$$

We have used the fact that partial derivatives commute. This term must vanish for the action to be invariant. Therefore it must be the case that

$$\partial\bar{\partial}X^\mu(z,\,\bar{z}) = 0 \tag{5.15}$$

We've written $X^\mu = X^\mu(z,\,\bar{z})$ to emphasize that in general the coordinates can be a function of z and $\bar{z}$. However, as you might guess from your studies of complex variables there is a special case of interest, that of analytic or *holomorphic* functions. A function $f(z,\,\bar{z})$ is holomorphic if

$$\frac{\partial f}{\partial \bar{z}} = 0 \tag{5.16}$$

That is, $f = f(z)$ only. On the other hand, if

$$\frac{\partial f}{\partial z} = 0 \tag{5.17}$$

and $f = f(\bar{z})$, then we say that f is *antiholomorphic*. In string theory, if $\bar{\partial}(\partial X^\mu) = 0$ then ∂X^μ is a holomorphic function which is called *left moving*. In the other case, where $\partial(\bar{\partial}X^\mu) = 0$, the function $\bar{\partial}X^\mu$ is antiholomorphic and is called *right moving*.

Generators of Conformal Transformations

To study the generators of a conformal transformation, we consider an infinitesimal transformation of the coordinates:

$$x'^{\mu} = x^{\mu} + \varepsilon^{\mu}$$

Now consider an infinitesimal conformal transformation. That is if $g'_{\mu\nu}(x') = \Omega(x)g_{\mu\nu}(x)$ we take $\Omega(x) = 1 - f(x)$ where $f(x)$ is some small departure from the identity. Then we have $g'_{\mu\nu}(x') = (1 - f(x))g_{\mu\nu}(x) = g_{\mu\nu}(x) - f(x)g_{\mu\nu}(x)$.

Using $x'^{\mu} = x^{\mu} + \varepsilon^{\mu}$ you can show that

$$g'_{\mu\nu} = g_{\mu\nu} - (\partial_{\mu}\varepsilon_{\nu} + \partial_{\nu}\varepsilon_{\mu})$$

So, recalling that we are working with a conformal transformation about the flat space metric, it must be the case that

$$\partial_{\mu}\varepsilon_{\nu} + \partial_{\nu}\varepsilon_{\mu} = f(x)g_{\mu\nu}$$

We can determine the form of $f(x)$ by multiplying both sides of this equation by $g^{\mu\nu}$. In d spacetime dimensions $g^{\mu\nu}g_{\mu\nu} = d$ and so on the right we obtain $g^{\mu\nu}f(x)g_{\mu\nu} = d\,f(x)$. On the left side we have

$$
\begin{aligned}
g^{\mu\nu}(\partial_{\mu}\varepsilon_{\nu} + \partial_{\nu}\varepsilon_{\mu}) &= g^{\mu\nu}\partial_{\mu}\varepsilon_{\nu} + g^{\mu\nu}\partial_{\nu}\varepsilon_{\mu} \\
&= \partial_{\mu}\varepsilon^{\mu} + \partial_{\nu}\varepsilon^{\nu} \quad \text{(raise indices with metric)} \\
&= \partial_{\mu}\varepsilon^{\mu} + \partial_{\mu}\varepsilon^{\mu} \quad \text{(relabel repeated indices} \\
&\qquad\qquad\qquad\qquad \text{which are dummy indices)} \\
&= 2\partial_{\mu}\varepsilon^{\mu}
\end{aligned}
$$

Hence

$$f = \frac{2}{d}\partial_{\mu}\varepsilon^{\mu}$$

And we have the relation

$$\partial_{\mu}\varepsilon_{\nu} + \partial_{\nu}\varepsilon_{\mu} = \frac{2}{d}\delta_{\mu\nu}\partial_{\rho}\varepsilon^{\rho} = \frac{2}{d}\delta_{\mu\nu}(\partial\cdot\varepsilon) \qquad (5.18)$$

Using $\square = \partial_\mu \partial^\mu$, taking the derivative with respect to ∂^μ we have, on the left-hand side:

$$\partial^\mu \partial_\mu \varepsilon_v + \partial^\mu \partial_v \varepsilon_\mu = \square \varepsilon_v + \partial_v (\partial \cdot \varepsilon)$$

Hence, taking the same derivative on the right side and equating results we obtain

$$\square \varepsilon_v + \left(1 - \frac{2}{d}\right) \partial_v (\partial \cdot \varepsilon) = 0$$

Notice that this equation singles out the case of two dimensions. Setting $d = 2$ we obtain

$$\square \varepsilon_v = 0$$

We can obtain a second equation which highlights the importance of $d = 2$ by operating on * with $\square = \partial_\mu \partial^\mu$. This gives

$$\{\delta_{\mu v} \square + (d - 2) \partial_\mu \partial_v\}(\partial \cdot \varepsilon) = 0$$

The infinitesimal parameter ε^μ can represent four different types of transformations: translations, scale transformations, rotations, and *special conformal transformations*. A translation takes the form

$$\varepsilon^\mu = a^\mu$$

where a^μ is a constant. A scale transformation is one of the form:

$$\varepsilon^\mu = \lambda x^\mu$$

For a rotation, we write

$$\varepsilon^\mu = \omega^\mu{}_v x^v$$

where we require that ω is antisymmetric, that is, $\omega_{\mu v} = -\omega_{v\mu}$. Finally, a special conformal transformation assumes the form

$$\varepsilon^\mu = b^\mu x^2 - 2x^\mu (b \cdot x)$$

These operations can be combined with the Poincaré group to form the *conformal group*. We incorporate two generators from the Poincaré group, the generator of translations P_μ and the generator of rotations $J_{\mu\nu}$. Denoting the generator of a scale transformation by D and the generator of a special conformal transformation by K_μ, the generators of the conformal group are

$$P_\mu = -i\partial_\mu \quad D = -ix\cdot\partial \quad J_{\mu\nu} = i(x_\mu\partial_\nu - x_\nu\partial_\mu) \quad K_\mu = -i[x^2\partial_\mu - 2x_\mu(x\cdot\partial)] \quad (5.19)$$

The Two-Dimensional Conformal Group

We now simplify the discussion somewhat and consider the special case of interest to us, the conformal group in two dimensions. In Eq. (5.18) we found that

$$\partial_\mu\varepsilon_\nu + \partial_\nu\varepsilon_\mu = \delta_{\mu\nu}\partial_\rho\varepsilon^\rho$$

where we have set $d = 2$. Proceeding with the two-dimensional case, take coordinates (x^1, x^2). Then when $\mu = 1$, $\nu = 2$ we have $\delta_{\mu\nu} = 0$ and we obtain

$$\partial_1\varepsilon_2 + \partial_2\varepsilon_1 = 0$$

Dispensing with the shorthand notation for a moment, this might look more familiar as

$$\frac{\partial\varepsilon_2}{\partial x^1} = -\frac{\partial\varepsilon_1}{\partial x^2}$$

This is nothing other than one of the Cauchy-Riemann equations when we take $\varepsilon = \varepsilon_1 + i\varepsilon_2$ and $x = x_1 + ix_2$. Similarly you can also show that

$$\frac{\partial\varepsilon_1}{\partial x^1} = \frac{\partial\varepsilon_2}{\partial x^2}$$

In the theory of complex variables we learned that a function that satisfies the Cauchy-Riemann equations in a given region R is called *analytic*. An analytic function is one that is a function of z only. So, labeling our coordinates with the usual complex coordinates $(z, \bar{z})$ conformal transformations in two dimensions are implemented using analytic functions:

$$z \to f(z) \quad \bar{z} \to \bar{f}(\bar{z}) \quad (5.20)$$

where $\bar{\partial} f = \partial \bar{f} = 0$. To obtain the generators, we consider a coordinate transformation of the form:

$$z \to z' = z - \varepsilon_n z^{n+1} \qquad \bar{z} \to \bar{z}' = \bar{z} - \bar{\varepsilon}_n \bar{z}^{n+1} \qquad (5.21)$$

To obtain an expression for the generators of a conformal transformation in two dimensions, we take the derivatives of the transformed coordinates z', $\bar{z}'$ and look for terms containing the derivatives $\partial \varepsilon_n$ and $\bar{\partial} \bar{\varepsilon}_n$, respectively. In the first case we obtain

$$\partial z' = \frac{\partial}{\partial z}(z - \varepsilon_n z^{n+1}) = 1 - \varepsilon_n (n+1) z^n - z^{n+1} \partial_z \varepsilon_n$$

This allows us to identify the generator:

$$\ell_n = -z^{n+1} \partial_z \qquad (5.22)$$

A similar procedure applied to the complex conjugate coordinate gives

$$\bar{\ell}_n = -\bar{z}^{n+1} \partial_{\bar{z}} \qquad (5.23)$$

In the classical case, the generators [Eqs. (5.22) and (5.23)] satisfy the Virasoro algebra:

$$[\ell_m, \ell_n] = (m-n)\ell_{m+n} \qquad [\bar{\ell}_m, \bar{\ell}_n] = (m-n)\bar{\ell}_{m+n} \qquad (5.24)$$

EXAMPLE 5.1

Show that the infinitesimal generator $\ell_n = -z^{n+1}\partial$ satisfies the Virasoro algebra $[\ell_m, \ell_n] = (m-n)\ell_{m+n}$.

SOLUTION

We apply the generator, which is an operator, to a test function f. So we obtain

$$
\begin{aligned}
[\ell_m, \ell_n]f &= (\ell_m \ell_n - \ell_n \ell_m)f \\
&= -z^{m+1}\partial(-z^{n+1}\partial)f - (-z^{n+1}\partial)(-z^{m+1}\partial)f \\
&= -z^{m+1}[-(n+1)z^n \partial f - z^{n+1}\partial^2 f] + z^{n+1}[-(m+1)z^m \partial f - z^{m+1}\partial^2 f] \\
&= (n+1)z^{m+n+1}\partial f + z^{m+n+2}\partial^2 f - (m+1)z^{m+n+1}\partial f - z^{m+n+2}\partial^2 f \\
&= (m-n)[-z^{m+n+1}\partial]f \\
&= (m-n)\ell_{m+n}f
\end{aligned}
$$

Hence we conclude that

$$[\ell_m, \ell_n] = (m-n)\ell_{m+n}$$

The generators $\ell_{0,\pm 1}$ and $\overline{\ell}_{0,\pm 1}$ are a special case. Consider the action of these generators on the infinitesimal coordinate transformations [Eq. (5.21)]. Taking $n = -1, 0, 1$ we have

$$n = -1 : z' = z - \varepsilon \quad \text{(translation)}$$
$$n = 0 : z' = z - \varepsilon z \quad \text{(scaling)}$$
$$n = 1 : z' = z - \varepsilon z^2 \quad \text{(special conformal transformation)}$$

There are similar expressions for the complex conjugates. Hence ℓ_{-1} and $\overline{\ell}_{-1}$ generate translation, ℓ_0 and $\overline{\ell}_0$, $(\ell_0 + \overline{\ell}_0)$ generate scaling and dilations, respectively, $i(\ell_0 - \overline{\ell}_0)$ generates rotations, and ℓ_1 and $\overline{\ell}_1$ generate special conformal transformations. All together, the transformations generated by $\ell_{0,\pm 1}$ and $\overline{\ell}_{0,\pm 1}$ can be written in the form

$$z \to \gamma(z) = \frac{az + b}{cz + d} \tag{5.25}$$

Here, $ad - bc = 1$ and the transformation $\gamma(z)$ is called a *Möbius transformation*.

EXAMPLE 5.2
Consider the transformation $T_a(z) = z / (1 + az)$. Show that $T_a(z)$ constitutes a transformation group by examining the composition $T_b(T_a(z))$.

SOLUTION
This is actually a simple problem. We need to show that

$$T_b(T_a(z)) = T_{a+b}(z) = \frac{z}{1 + (a+b)z}$$

Starting with $T_a(z) = z / (1 + az)$, let $w = z / (1 + az)$. Now

$$T_b(w) = \frac{w}{1 + bw}$$

Using $w = z/(1 + az)$ we get

$$T_b(w) = \frac{w}{1+bw} = \frac{\dfrac{z}{1+az}}{1+b\left(\dfrac{z}{1+az}\right)}$$

$$= \frac{z}{(1+az)\left(1+b\left(\dfrac{z}{1+az}\right)\right)}$$

$$= \frac{z}{1+az+bz} = \frac{z}{1+(a+b)z} = T_{a+b}(z)$$

Hence, $T_a(z) = z/(1+az)$ satisfies the group composition property.

EXAMPLE 5.3
Let $T_a(z) = z/(1+az)$ and suppose that a is real and $|a| = 1$. Determine the generators of this transformation.

SOLUTION
Recall that the generators have the form $\ell_n = -z^{n+1}\partial_z$ and similarly for the complex conjugate. This form holds for an infinitesimal transformation parameter $\varepsilon(z) = -\sum a_n z^{n+1}$. So we can deduce the expressions for the generators by writing the transformation as a series.

Consider the following series

$$s = 1 - r + r^2 - r^3 + \cdots$$

Now multiply by r to give

$$rs = r - r^2 + r^3 - r^4 + \cdots$$

Now add both series. On the left side we obtain $s + rs = s(1+r)$. On the right, we have

$$s + rs = 1 - r + r^2 - r^3 + \cdots + r - r^2 + r^3 - r^4 + \cdots = 1$$

This tells us that

$$\frac{1}{1+r} = s = 1 - r + r^2 - r^3 + \cdots$$

So, taking into account that $|a| = 1$ we can write the transformation as

$$T_a(z) = \frac{z}{1+az} = z(1 - az + O(a^2)) \approx z - az^2 + O(a^2)$$

The power associated with the small parameter is z^2 from which we deduce that $n = 1$. So the generator is

$$\ell_1 = -z^2 \partial_z$$

We aren't quite done—we need to consider the complex conjugate. This takes the same form:

$$\overline{T_a(z)} = \frac{\overline{z}}{1+\overline{az}} = \overline{z}(1 - a\overline{z} + O(a^2)) \approx \overline{z} - a\overline{z}^2 + O(a^2)$$

Notice we used the fact that a is real. The generator in this case is $\overline{\ell}_1 = -\overline{z}^2 \partial_{\overline{z}}$, and the generator for the transformation is found by taking the sum:

$$\ell_1 + \overline{\ell}_1 = -z^2 \partial_z - \overline{z}^2 \partial_{\overline{z}}$$

Central Extension

In the quantum theory the Virasoro operators acquire an extra term which goes by the name *central extension*. Calling c the *central charge*, the algebra described by Eq. (5.24) becomes

$$[L_m, L_n] = (m - n)L_{m+n} + \frac{c}{12} m(m^2 - 1)\delta_{m+n,0} \qquad (5.26)$$

The result in Eq. (5.26) is known as the *Virasoro algebra*. The classical form [Eq. (5.24)] is sometimes called the *Witten algebra*. This formula [Eq. (5.26)] will be derived in the next chapter using the operator product expansion of the energy-momentum tensor.

Closed String Conformal Field Theory

Now let's move forward so we can see what the developments laid out thus far in the chapter really mean. We will see that the energy-momentum tensor can be expanded in a Laurent series, and that the Virasoro operators turn out to be the coefficients of the expansion. In other words, they describe the modes of the energy-momentum tensor. In particular, the operator L_0 is proportional to the energy operator or hamiltonian.

As a specific example, consider a closed string with worldsheet coordinates (τ, σ). The spatial dimension is compactified, that it is periodic with

$$\sigma = \sigma + 2\pi \tag{5.27}$$

The time coordinate satisfies $-\infty < \tau < \infty$. We can describe the worldsheet of the closed string, which is an infinite cylinder, using conformal field theory in the following way. We begin by making the following conformal transformation:

$$z = e^{\tau + i\sigma} \qquad \overline{z} = e^{\tau - i\sigma} \tag{5.28}$$

The effect of this transformation is to map the cylinder to the complex plane. The radial coordinate plays the role of time, with the infinite past at the origin. With increasing radius, we move forward in time. Spatial integrals on the worldsheet are translated into contour integrals about the origin in the complex plane as the result of the conformal transformation [Eq. (5.28)]. A slice through the cylinder, which corresponds to a slice at constant time τ_i, is transformed into a circle of radius r_i in the complex plane. That is, radius in the z plane is a measure of Euclidean worldsheet time as

$$R = |z| = e^{\tau}$$

So, at time τ_1 a closed string is a circle of radius $R_1 = |z| = e^{\tau_1}$ in the z plane, with the angular coordinate given by $\theta = \sigma$. This is illustrated in Fig. 5.1.

As time increases, from say $\tau_1 \to \tau_2$, $\tau_2 > \tau_1$, the radius of the circle in the z plane increases from R_1 to $R_2 > R_1$.

Let's recall the left-moving and right-moving modes described in Eq. (2.55) and (2.56). Given Eq. (5.28), it is clear that $\tau + \sigma \infty \ln z$, and $\tau - \sigma \infty \ln \overline{z}$. So we have

$$X_L^\mu(\overline{z}) = \frac{x^\mu}{2} - i\frac{\ell_s^2}{2}p^\mu \ln \overline{z} + i\frac{\ell_s}{\sqrt{2}}\sum_{n\neq 0}\frac{\tilde{\alpha}_n^\mu}{n}\overline{z}^{-n} \tag{5.29}$$

$$X_R^\mu(z) = \frac{x^\mu}{2} - i\frac{\ell_s^2}{2}p^\mu \ln z + i\frac{\ell_s}{\sqrt{2}}\sum_{n\neq 0}\frac{\alpha_n^\mu}{n}z^{-n} \tag{5.30}$$

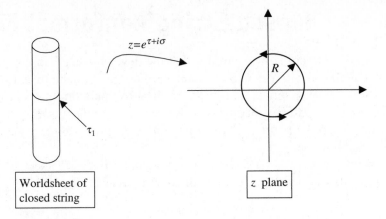

Worldsheet of
closed string

z plane

Figure 5.1. The worldsheet of a closed string is mapped to the z plane. A slice
through the cylinder, at a constant time, is mapped to a circle of a
fixed radius in the z plane. The radius of the circle in the z plane
corresponds to (Euclidean) time on the worldsheet.

The energy-momentum tensor of the worldsheet is a conserved quantity, meaning
that

$$\partial^\mu T_{\mu\nu} = 0 \tag{5.31}$$

The energy-momentum tensor is also traceless. This means that in the coordinates
$(z, \overline{z})$, the components $T_{z\overline{z}} = 0$. The conservation condition [Eq. (5.31)] implies that

$$\partial_z T_{\overline{z}\overline{z}} = \partial_{\overline{z}} T_{zz} = 0 \tag{5.32}$$

That is, the energy-momentum tensor is composed of a holomorphic and
antiholomorphic functions given by T_{zz} and $T_{\overline{z}\overline{z}}$, respectively. A holomorphic
function has a Laurent series expansion, which we write as

$$T_{zz}(z) = \sum_{m=-\infty}^{\infty} \frac{L_m}{z^{m+2}} \tag{5.33}$$

We have written this expression in a way anticipating that the Laurent coefficients
are the Viarasoro generators. The antiholomorphic component also has a Laurent
expansion:

$$T_{\overline{z}\overline{z}}(\overline{z}) = \sum_{m=-\infty}^{\infty} \frac{\overline{L}_m}{\overline{z}^{m+2}} \tag{5.34}$$

Using standard complex analysis, this tells us that we can invert these formulas in the following way:

$$L_m = \frac{1}{2\pi i} \oint \frac{dz}{z} z^{m+2} T_{zz}(z) \qquad \bar{L}_m = \frac{1}{2\pi i} \oint \frac{d\bar{z}}{\bar{z}} \bar{z}^{m+2} T_{\bar{z}\bar{z}}(\bar{z}) \qquad (5.35)$$

The deformation of path theorem (see *Complex Variables Demystified*) tells us that we can shrink or expand the contour used for the integration in Eq. (5.35) and leave the integral invariant. Since movement along the radial distance in the complex plane corresponds to time translation, this tells us that the Virasoro operators are invariant under a time translation, in other words the integrals [Eq. (5.35)] are related to conserved charges.

Wick Expansion

To gain insight into the physics of the problem, we will begin by calculating propagators for the theory. In the text we will derive the closed string *propagator*, you can try the open string case in the chapter quiz.

We begin by considering the propagator or Wick expansion for $X^{\mu}(\sigma, \tau)$ for the closed string. This is done by calculating

$$\left\langle X^{\mu}(\sigma, \tau), X^{\nu}(\sigma', \tau') \right\rangle = T(X^{\mu}(\sigma, \tau), X^{\nu}(\sigma', \tau')) - : X^{\mu}(\sigma, \tau) X^{\nu}(\sigma', \tau') :$$

This is also called the *two-point function*. There are two notations you should be aware of in this expression. The T indicates that the expression is *time ordered*. It is shorthand for

$$T(X^{\mu}(\sigma, \tau), X^{\nu}(\sigma', \tau')) = X^{\mu}(\sigma, \tau) X^{\nu}(\sigma', \tau') \theta(\tau - \tau') + X^{\nu}(\sigma', \tau') X^{\mu}(\sigma, \tau) \theta(\tau' - \tau)$$

To understand what we have here, note that the Heaviside function $\theta(t) = 1$ if $t > 0$, $\theta(t) = 0$ if $t < 0$. This expression ensures that the term which occurs earlier in time is to the right. So if $\tau > \tau'$, which means that τ is later than τ', then

$$T(X^{\mu}(\sigma, \tau), X^{\nu}(\sigma', \tau')) = X^{\mu}(\sigma, \tau) X^{\nu}(\sigma', \tau')$$

and vice versa for the other possibility. The colons indicate *normal ordering* which puts all creating operators to the left of all annihilation operators. Now consider the fact that we can write the fields in terms of left movers and right movers. Then

$$\langle 0|X^\mu(\sigma, \tau)X^\nu(\sigma', \tau')|0\rangle = \langle 0|(X_L^\mu + X_R^\mu)(X_L^\nu + X_R^\nu)|0\rangle$$

$$= \langle 0|X_L^\mu X_L^\nu + X_L^\mu X_R^\nu + X_R^\mu X_L^\nu + X_R^\mu X_R^\nu|0\rangle$$

Now let's move to the complex plane where $X_R^\mu = X_R^\mu(z)$, $X_L^\mu = X_L^\mu(\bar{z})$ so that $\langle 0|X^\mu(\sigma, \tau)X^\nu(\sigma', \tau')|0\rangle \to \langle 0|X^\mu(z, \bar{z})X^\nu(z', \bar{z}')|0\rangle$. Let's consider one term, $\langle 0|X_R^\mu X_R^\nu|0\rangle$ using the expansion given in Eq. (5.30). Noting that $p_\mu|0\rangle = 0$ and $[x^\mu, p^\nu] = i\eta^{\mu\nu}$, $\alpha_m^\mu|0\rangle = 0$ if $m > 0$ and $\langle 0|\alpha_n^\mu = 0$ for $n < 0$, we have

$$\langle 0|X_R^\mu X_R^\nu|0\rangle = \frac{1}{4}\langle 0|x^\mu x^\nu|0\rangle - i\frac{\ell_s^2}{4}\ln z\langle 0|p^\mu x^\nu|0\rangle - \frac{\ell_s^2}{2}\sum_{\substack{m\neq 0 \\ n\neq 0}}\frac{1}{mn}\langle 0|\alpha_n^\mu \alpha_m^\nu|0\rangle z^{-n}z'^{-m}$$

$$= \frac{1}{4}\langle 0|x^\mu x^\nu|0\rangle - \frac{\eta^{\mu\nu}\ell_s^2}{2}\frac{\ln z}{2} + \frac{\ell_s^2}{2}\sum_{\substack{m\neq 0 \\ n\neq 0}}\frac{1}{mn}\langle 0|\alpha_n^\mu \alpha_{-m}^\nu|0\rangle z^{-n}z'^{-m}$$

Moving from the first to the second line, we used

$$\langle 0|p^\mu x^\nu|0\rangle = \langle 0|x^\nu p^\mu - i\eta^{\mu\nu}|0\rangle = \langle 0|x^\nu p^\mu|0\rangle - i\eta^{\mu\nu}\langle 0|0\rangle = -i\eta^{\mu\nu}$$

Now the commutator of the creation and annihilation operators satisfies

$$\left[\alpha_m^\mu, \alpha_n^\nu\right] = m\eta^{\mu\nu}\delta_{m,-n}$$

And so this becomes

$$\langle 0|X_R^\mu X_R^\nu|0\rangle = \frac{1}{4}\langle 0|x^\mu x^\nu|0\rangle - \frac{\eta^{\mu\nu}\ell_s^2}{2}\frac{\ln z}{2} + \frac{\ell_s^2}{2}\sum_{n>0}\frac{1}{n}\eta^{\mu\nu}z^{-n}z'^n$$

Now, using the fact that $\sum_{n=1}^{\infty} x^n/n = -\ln(1-x)$, we can write this as

$$\langle 0|X_R^{\mu} X_R^{\nu}|0\rangle = \frac{1}{4}\langle 0|x^{\mu}x^{\nu}|0\rangle - \frac{\eta^{\mu\nu}\ell_s^2}{4} + \frac{\ell_s^2}{2}\eta^{\mu\nu}\sum_{n>0}\frac{(z'/z)^n}{n}$$

$$= \frac{1}{4}\langle 0|x^{\mu}x^{\nu}|0\rangle - \frac{\eta^{\mu\nu}\ell_s^2}{2}\left(\frac{\ln z}{2} + \ln(1 - z'/z)\right)$$

$$= \frac{1}{4}\langle 0|x^{\mu}x^{\nu}|0\rangle + \frac{\eta^{\mu\nu}\ell_s^2}{4}\ln z - \frac{\eta^{\mu\nu}\ell_s^2}{2}\ln(z - z')$$

Similar calculations show that

$$\langle 0|X_L^{\mu} X_L^{\nu}|0\rangle = \frac{1}{4}\langle 0|x^{\mu}x^{\nu}|0\rangle + \frac{\eta^{\mu\nu}\ell_s^2}{4}\ln\bar{z} - \frac{\eta^{\mu\nu}\ell_s^2}{2}\ln(\bar{z} - \bar{z}')$$

$$\langle 0|X_R^{\mu} X_L^{\nu}|0\rangle = -\frac{\eta^{\mu\nu}\ell_s^2}{4}\ln z$$

$$\langle 0|X_L^{\mu} X_R^{\nu}|0\rangle = -\frac{\eta^{\mu\nu}\ell_s^2}{4}\ln\bar{z}$$

Adding these terms up we find that

$$\langle 0|X^{\mu}(z, \bar{z})X^{\nu}(z', \bar{z}')|0\rangle$$

$$= \langle 0|x^{\mu}x^{\nu}|0\rangle - \frac{\eta^{\mu\nu}\ell_s^2}{2}\ln[(z - z')(\bar{z} - \bar{z}')] = \langle 0|X^{\nu}(z', \bar{z}')X^{\mu}(z, \bar{z})|0\rangle$$

So the vacuum expectation value of the time ordered product is

$$\langle 0|T[X^{\mu}(z, \bar{z})X^{\nu}(z', \bar{z}')]|0\rangle = \langle 0|x^{\mu}x^{\nu}|0\rangle - \frac{\eta^{\mu\nu}\ell_s^2}{2}\ln[(z - z')(\bar{z} - \bar{z}')]$$

Now let's return to the two-point function:

$$\left\langle X^{\mu}(\sigma, \tau), X^{\nu}(\sigma', \tau')\right\rangle = T(X^{\mu}(\sigma, \tau), X^{\nu}(\sigma', \tau')) - : X^{\mu}(\sigma, \tau)X^{\nu}(\sigma', \tau') :$$

Normal ordering puts all destruction operators to the right. So $: p^{\mu}x^{\nu} : |0\rangle = x^{\nu}p^{\mu}|0\rangle = 0$. Also, normal ordering of terms like $\sum_{\substack{m \neq 0 \\ n \neq 0}} 1/(mn)\langle 0|\alpha_n^{\mu}\alpha_m^{\nu}|0\rangle z^{-n}z'^{-m}$

puts all destruction operators to the right annihilating the vacuum. So the vacuum expectation value of the normal ordered piece reduces to

$$\langle 0 | : X^\mu(\sigma, \tau) X^\nu(\sigma', \tau') : | 0 \rangle = \langle 0 | x^\mu x^\nu | 0 \rangle$$

Hence the two-point function is

$$\langle X^\mu(\sigma, \tau), X^\nu(\sigma', \tau') \rangle = T(X^\mu(\sigma, \tau), X^\nu(\sigma', \tau')) - : X^\mu(\sigma, \tau) X^\nu(\sigma', \tau') :$$

$$= -\frac{\eta^{\mu\nu} \ell_s^2}{2} \ln\left[(z - z')(\bar{z} - \bar{z}') \right]$$

$$= -\frac{\eta^{\mu\nu} \ell_s^2}{2} \ln(z - z') - \frac{\eta^{\mu\nu} \ell_s^2}{2} \ln(\bar{z} - \bar{z}')$$

Now that we have this expression, we can easily calculate other quantities that involve derivatives, say. For example, to calculate the two-point function $\langle \partial_z X^\mu(z, \bar{z}), X^\nu(z', \bar{z}') \rangle$, we just differentiate the result:

$$\langle \partial_z X^\mu(z, \bar{z}), X^\nu(z', \bar{z}') \rangle = \partial_z \left[-\frac{\eta^{\mu\nu} \ell_s^2}{2} \ln(z - z') - \frac{\eta^{\mu\nu} \ell_s^2}{2} \ln(\bar{z} - \bar{z}') \right]$$

$$= -\frac{\eta^{\mu\nu} \ell_s^2}{2} \frac{1}{z - z'}$$

And

$$\langle \partial_z X^\mu(z, \bar{z}), \partial_{z'} X^\nu(z', \bar{z}') \rangle = \frac{\eta^{\mu\nu} \ell_s^2}{2} \frac{1}{(z - z')^2}$$

Operator Product Expansion

A key concept we need to continue forward is known as an *operator product expansion*. This is often given the abbreviation *OPE*. An operator product expansion is a series expansion of a product of two operator-valued fields. Let's denote these fields by A_i and consider two space-time points z and w. Then in a region R that does not contain w

$$A_i(z) A_j(w) = \sum_k c_{ijk}(z - w) A_k(w) \tag{5.36}$$

The $c_{ijk}(z-w)$ are analytic functions in R and the $A_k(w)$ are operator-valued fields. Now, define a conformal transformation $z \to w(z)$. A *conformal field* or *primary field* is one that transforms as

$$\Phi(z, \bar{z}) = \left(\frac{\partial w}{\partial z} \right)^h \left(\frac{\partial \bar{w}}{\partial \bar{z}} \right)^{\bar{h}} \Phi(w, \bar{w}) \qquad (5.37)$$

We say that $(h, \bar{h})$ are the *conformal weights* or *conformal dimension* of the field. In particular, $h + \bar{h}$ is the dimension which describes how the field Φ behaves under scaling, while $h - \bar{h}$ is the *spin* of Φ, which describes how the field is transformed under a rotation. If Eq. (5.37) is satisfied, then it follows that

$$\Phi(z, \bar{z})(dz)^h (d\bar{z})^{\bar{h}} = \Phi(w, \bar{w})(dw)^h (d\bar{w})^{\bar{h}} \qquad (5.38)$$

That is, $\Phi(z, \bar{z})(dz)^h (d\bar{z})^{\bar{h}}$ is invariant under a conformal transformation.

Working in the complex plane, time ordering is transformed into *radial ordering* because as we mentioned above, the radial direction encodes the flow of time in the z plane. Consider two operators defined in the complex plane $A(z)$ and $B(w)$. The radial-ordering operator R fixes the order of the operators based on which one has the larger radius in the complex plane. That is,

$$R[A(z)B(w)] = \begin{cases} A(z)B(w), & |z| > |w| \\ (-1)^f B(w)A(z), & |w| > |z| \end{cases}$$

If the operators are fermionic, then $f = 1$.

One operator product expansion of particular interest involves the energy-momentum tensor. In the complex plane

$$T_{zz}(z) =: \eta_{\mu\nu} \partial_z X^\mu \partial_z X^\nu : \qquad (5.39)$$

EXAMPLE 5.4
Find the operator product expansion of the radially ordered product $T_{zz}(z)\partial_w X^\rho(w)$.

SOLUTION
Using Eq. (5.39) we have

$$\left\langle R(T_{zz}(z)\partial_w X^\rho(w)) \right\rangle = R(: \eta_{\mu\nu} \partial_z X^\mu(z) \partial_z X^\nu(z) : \partial_w X^\rho(w))$$

$$= \eta_{\mu\nu} \left\langle \partial_z X^\mu(z) \partial_w X^\rho(w) \right\rangle \partial_z X^\nu(z)$$

$$+ \eta_{\mu\nu} \left\langle \partial_z X^\nu(z) \partial_w X^\rho(w) \right\rangle \partial_z X^\mu(z)$$

Now we can use our previous result, namely

$$\langle \partial_z X^\mu(z, \bar{z}), \partial_{z'} X^\nu(z', \bar{z}') \rangle = \frac{\eta^{\mu\nu} \ell_s^2}{2} \frac{1}{(z-z')^2}$$

And so we obtain

$$\langle R(T_{zz}(z) \partial_w X^\rho(w)) \rangle$$

$$= -\frac{\ell_s^2}{2} \frac{1}{(z-w)^2} \eta_{\mu\nu} \eta^{\mu\rho} \partial_z X^\nu(z) - \frac{\ell_s^2}{2} \frac{1}{(z-w)^2} \eta_{\mu\nu} \eta^{\nu\rho} \partial_z X^\mu(z)$$

$$= -\ell_s^2 \frac{1}{(z-w)^2} \partial_z X^\rho(z)$$

Now expand $\partial_z X^\rho(z)$ in a power series about the point $z = w$. We find

$$\partial_z X^\rho(z) = \partial_w X^\rho(w) + (z-w) \partial_z^2 X^\rho(z) + \frac{(z-w)^2}{2!} \partial_z^3 X^\rho(z) + \cdots$$

Hence

$$\langle R(T_{zz}(z) \partial_w X^\rho(w)) \rangle$$

$$= -\ell_s^2 \frac{1}{(z-w)^2} \partial_z X^\rho(z)$$

$$= -\ell_s^2 \frac{1}{(z-w)^2} \left\{ \partial_w X^\rho(w) + (z-w) \partial_z^2 X^\rho(z) + \frac{(z-w)^2}{2!} \partial_z^3 X^\rho(z) + \cdots \right\}$$

$$= -\ell_s^2 \frac{1}{(z-w)^2} \partial_w X^\rho(w) - \frac{\ell_s^2}{z-w} \partial_z^2 X^\rho(z) - \ell_s^2 \frac{(z-w)}{2!} \partial_z^3 X^\rho(z) + \cdots$$

The *singular* terms are what is of interest. So it is typical to use an ellipsis $\cdots$ to represent the regular terms in the summation and write

$$\langle R(T_{zz}(z) \partial_w X^\rho(w)) \rangle = -\ell_s^2 \frac{1}{(z-w)^2} \partial_w X^\rho(w) - \frac{\ell_s^2}{z-w} \partial_z^2 X^\rho(z) + \cdots$$

EXAMPLE 5.5

In this example we compute an important result, the OPE of $T_{zz}(z)T_{ww}(w)$.

SOLUTION

Using radial ordering we have

$$R(T_{zz}(z)T_{ww}(w)) = R(\eta_{\mu\nu} : \partial_z X^\mu(z)\partial_z X^\nu(z) : \eta_{\rho\sigma} : \partial_w X^\rho(w)\partial_w X^\sigma(w) :)$$

$$= 2\eta_{\mu\nu}\eta_{\rho\sigma} \langle \partial_z X^\mu(z)\partial_w X^\rho(w)\rangle\langle \partial_z X^\nu(z)\partial_w X^\sigma(w)\rangle$$

$$+ 4\eta_{\mu\nu}\eta_{\rho\sigma} \langle \partial_z X^\mu(z)\partial_w X^\rho(w)\rangle : \partial_z X^\nu(z)\partial_w X^\sigma(w) :$$

We have already seen that

$$\langle \partial_z X^\mu(z),\ \partial_w X^\rho(w)\rangle = -\frac{\eta^{\mu\rho}\ell_s^2}{2}\frac{1}{(z-w)^2}$$

Hence

$$\langle \partial_z X^\mu(z)\partial_w X^\rho(w)\rangle\langle \partial_z X^\nu(z)\partial_w X^\sigma(w)\rangle$$

$$= \left(\frac{\eta^{\mu\rho}\ell_s^2}{2}\frac{1}{(z-w)^2}\right)\left(\frac{\eta^{\nu\sigma}\ell_s^2}{2}\frac{1}{(z-w)^2}\right) = \frac{\eta^{\mu\rho}\eta^{\nu\sigma}\ell_s^4}{4}\frac{1}{(z-w)^4}$$

Now the metric terms give us the dimension of the space:

$$\eta_{\mu\nu}\eta_{\rho\sigma}\eta^{\mu\rho}\eta^{\nu\sigma} = \delta^\rho_\nu\delta^\nu_\rho = D$$

In the last term of $R(T_{zz}(z)T_{ww}(w))$ which includes the normal ordered product, we expand $\partial_z X^\nu(z)$ about w the way we did in Example 5.4. Then we obtain the operator product expansion for the energy-momentum tensor:

$$R(T_{zz}(z)T_{ww}(w)) = 2\eta_{\mu\nu}\eta_{\rho\sigma} \langle \partial_z X^\mu(z)\partial_w X^\rho(w)\rangle\langle \partial_z X^\nu(z)\partial_w X^\sigma(w)\rangle$$

$$+ 4\eta_{\mu\nu}\eta_{\rho\sigma} \langle \partial_z X^\mu(z)\partial_w X^\rho(w)\rangle : \partial_z X^\nu(z)\partial_w X^\sigma(w) :$$

$$= \ell_s^2 \frac{D/2}{(z-w)^4} - \ell_s^2 \frac{2}{(z-w)^2} T_{ww}(w) - \ell_s^2 \frac{1}{z-w}\partial_w T_{ww}(w) + \cdots$$

Summary

Conformal field theory is a tool that allows us to transform the physics of the string worldsheet to the complex plane. Calculations are much easier to do using the techniques of complex variables. In this chapter we introduced some of the basic terminology and techniques. In the following chapters, we will continue our exploration of conformal field theory in the context of string theory by exploring Kac-Moody algebras, minimal models, vertex operators, and BRST quantization.

Quiz

1. Calculate $[\ell_m, \overline{\ell}_n]$.

2. Does $T_a(z) = \dfrac{1}{1+az}$ satisfy the group composition property?

3. Consider $T_a(z) = \dfrac{z}{1+az}$ with a pure imaginary. Find the generator of the transformation.

4. Following the text, calculate $\langle 0 | X_L^\mu X_L^\nu | 0 \rangle$.

5. For a closed string, calculate $\langle \partial_{\overline{z}} X^\mu(z, \overline{z}), \partial_{\overline{z}'} X^\nu(z', \overline{z}') \rangle$.

6. Calculate $\langle 0 | : X^\mu(z, \overline{z}) X^\nu(z', \overline{z}') : | 0 \rangle$.

7. Find $\langle 0 | X^\mu(z, \overline{z}) X^\nu(z', \overline{z}') | 0 \rangle$.

8. By exploiting the properties of the natural logarithm function, and using the fact that $\langle 0 | X^\mu(z, \overline{z}) X^\nu(z', \overline{z}') | 0 \rangle = \langle 0 | X^\nu(z', \overline{z}') X^\mu(z, \overline{z}) | 0 \rangle$, find a compact expression for $\langle 0 | X^\mu(z, \overline{z}) X^\nu(z', \overline{z}') | 0 \rangle$.

9. Find the operator product expansion of $R(T_{zz}(z) : e^{ik \cdot X(w)} :)$.

CHAPTER 6

BRST Quantization

So far we have seen two methods that can be utilized to quantize the string: the covariant approach and light-cone quantization. Each offers its advantages. Covariant quantization makes Lorentz invariance manifest but allows for the existence of "ghost states" (states with negative norm) in the theory. In contrast, light-cone quantization is ghost free. However, Lorentz invariance is no longer obvious. Another trade-off is that the proof of the number of space-time dimensions ($D = 26$ for the bosonic theory) is rather difficult in covariant quantization, but it's rather straightforward in light-cone quantization. Finally identifying the physical states is easier in the light-cone approach.

Another method of quantization, that in some ways is a more advanced approach, is called *BRST quantization*. This approach takes a middle ground between the two methods outlined above. BRST quantization is manifestly Lorentz invariant, but includes ghost states in the theory. Despite this, BRST quantization makes it easier to identify the physical states of the theory and to extract the number of space-time dimensions relatively easily.

BRST Operators and Introductory Remarks

We begin by considering our old friend, *Lie algebra*. This is the algebra that is obeyed by the familiar spin angular momentum operators of ordinary quantum mechanics. In general, let some physical theory contain a gauge symmetry with operators K_i. These operators satisfy the Lie algebra:

$$[K_i, K_j] = f_{ij}{}^k K_k \tag{6.1}$$

where the $f_{ij}{}^k$ are called the *structure constants* of the theory (note we are using the Einstein summation convention, so repeated indices are summed over). The structure constants satisfy

$$f_{ij}{}^m f_{mk}{}^l + f_{jk}{}^m f_{mi}{}^l + f_{ki}{}^m f_{mj}{}^l = 0 \tag{6.2}$$

The BRST quantization procedure begins with the following. We introduce two *ghost fields* that are denoted by b_i and c_j which satisfy an *anticommutation relation* given by

$$\{c^i, b_j\} = \delta^i_j \tag{6.3}$$

Furthermore $\{c^i, c^j\} = \{b_i, b_j\} = 0$. Here the c^i are ghost fields and the b_j are "ghost momenta."

Notice that since an anticommutation relation is satisfied by the ghost fields, these fields are *fermionic*. Now, recall that a field $\phi(z, \bar{z})$ has conformal dimension $(h, \bar{h})$ provided that it transforms under some conformal transformation $z \to w(z)$ as follows:

$$\phi(z, \bar{z}) = \left(\frac{\partial w}{\partial z}\right)^h \left(\frac{\partial w}{\partial \bar{z}}\right)^{\bar{h}} \varphi(w, \bar{w}) \tag{6.4}$$

The fields b and c are chosen such that they have conformal dimension 2 and -1, as we will see later.

There are two operators that are constructed out of the ghost fields and the K_i. The first of these is the *BRST operator* which is given by

$$Q = c^i K_i - \frac{1}{2} f_{ij}{}^k c^i c^j b_k \tag{6.5}$$

It is assumed that $Q = Q^\dagger$. We say that the BRST operator is *nilpotent* of degree two. This means that if we square the operator we get zero:

$$Q^2 = 0 \tag{6.6}$$

Notice that Eq. (6.6) can also be expressed as $\{Q, Q\} = 0$. We label the BRST operator with a Q to imply that this is a conserved charge of the system, we often call this the *BRST charge*. A second operator that is composed solely of ghost fields is called the *ghost number operator U*. This is given by

$$U = c^i b_i \tag{6.7}$$

(Again note the Einstein summation convention is being used.) This operator has integer eigenvalues. If the dimension of the Lie algebra is n, then the eigenvalues of U are the integers $0,...,n$. A state $|\psi\rangle$ has ghost number m if $U|\psi\rangle = m|\psi\rangle$.

EXAMPLE 6.1
Show that Q raises the ghost number by 1.

SOLUTION
Using the anticommutation relations for the ghost fields [Eq. (6.3)], notice that

$$Uc^i K_i = \sum_r c^r b_r c^i K_i = \sum_r c^r \left(\delta_r^i - c^i b_r\right) K_i$$

$$= c^i K_i - c^i \sum_r c^r b_r K_i$$

$$= c^i K_i + c^i K_i \sum_r c^r b_r = c^i K_i + c^i K_i U$$

Now consider some state $|\psi\rangle$ with ghost number m, that is $U|\psi\rangle = m|\psi\rangle$. Then

$$U\left(Q|\psi\rangle\right) = U\left(c^i K_i - \frac{1}{2} f_{ij}^{\ k} c^i c^j b_k\right)|\psi\rangle$$

$$= \left(Uc^i K_i - \left(\sum_r c^r b_r\right)\frac{1}{2} f_{ij}^{\ k} c^i c^j b_k\right)|\psi\rangle$$

$$= \left(c^i K_i + c^i K_i U - \frac{1}{2} f_{ij}^{\ k} \sum_r c^r b_r c^i c^j b_k\right)|\psi\rangle$$

$$\left(\text{Use } Uc^i K_i = c^i K_i + c^i K_i U\right)$$

$$= \left(c^i K_i + c^i K_i U - \frac{1}{2} f_{ij}^{\ k} \sum_r c^r \left(\delta_r^i - c^i b_r \right) c^j b_k \right) |\psi\rangle \qquad \text{[Apply Eq. (6.3)]}$$

$$= \left(c^i K_i + c^i K_i U - \frac{1}{2} f_{ij}^{\ k} c^i c^j b_k + \frac{1}{2} f_{ij}^{\ k} c^i \sum_r c^r b_r c^j b_k \right) |\psi\rangle$$

(Use the properties of δ_r^i)

$$= \left(c^i K_i + c^i K_i U - \frac{1}{2} f_{ij}^{\ k} c^i c^j b_k + \frac{1}{2} f_{ij}^{\ k} c^i \sum_r c^r \left(\delta_r^j - c^j b_r \right) b_k \right) |\psi\rangle$$

[Apply Eq. (6.3) again]

$$= \left(Q + c^i K_i U + \frac{1}{2} f_{ij}^{\ k} c^i \sum_r c^r \left(\delta_r^j - c^j b_r \right) b_k \right) |\psi\rangle \qquad \text{[Use Eq. (6.5) to write } Q\text{]}$$

$$= \left(Q + c^i K_i U + \frac{1}{2} f_{ij}^{\ k} c^i c^j b_k - \frac{1}{2} f_{ij}^{\ k} c^i c^j \sum_r c^r b_k b_r \right) |\psi\rangle$$

(Kroneker delta again)

$$= \left(Q + c^i K_i U + \frac{1}{2} f_{ij}^{\ k} c^i c^j b_k - \frac{1}{2} f_{ij}^{\ k} c^i c^j \sum_r \left(\delta_r^k - b_k c^r \right) b_r \right) |\psi\rangle$$

[Yet again use Eq. (6.3)]

$$= \left(Q + c^i K_i U - \frac{1}{2} f_{ij}^{\ k} c^i c^j \sum_r c^r b_r \right) |\psi\rangle$$

$$= (Q + QU)|\psi\rangle = (1 + m)(Q|\psi\rangle) \qquad \left(\text{Apply } U|\psi\rangle = m|\psi\rangle \right)$$

Hence the BRST operator raises the ghost number by 1.

BRST-Invariant States

A state $|\psi\rangle$ is called *BRST invariant* if it is annihilated by the BRST operator [Eq. (6.5)]:

$$Q|\psi\rangle = 0 \Rightarrow |\psi\rangle \text{ is BRST invariant} \qquad (6.8)$$

BRST-invariant states are the *physical states of the theory.* Since $Q^2 = 0$, it follows that any state $|\psi\rangle = Q|\chi\rangle \neq 0$ is BRST invariant, since

$$Q|\psi\rangle = Q^2|\chi\rangle = 0$$

Call $|\psi\rangle = Q|\chi\rangle$ a *null state*. Suppose that $|\phi\rangle$ is an arbitrary physical state so that $Q|\phi\rangle = 0$. Notice that

$$\langle\phi|\psi\rangle = \langle\phi|(Q|\chi\rangle) = ((\langle\phi|Q)|\chi\rangle = 0$$

Hence, any amplitude taken between a physical state and a null state vanishes. This implies a result, which is somewhat akin to the idea of a phase factor in ordinary quantum mechanics. Two states $|a\rangle, |b\rangle = e^{i\theta}|a\rangle$ represent the same physical state in quantum mechanics since the resulting amplitudes are the same. Likewise, since any inner product between a physical state and a null state vanishes, adding a null state $|\psi\rangle = Q|\chi\rangle$ to a physical state $|\phi\rangle$ generates a new state which is physically equivalent to $|\phi\rangle$:

$$|\phi'\rangle = |\phi\rangle + Q|\chi\rangle$$

because the second term on the right does not contribute to any inner product and hence does not change the physical predictions of the theory, which are the amplitudes. Since Q raises the ghost number of a state by 1, if the ghost number of $|\psi\rangle$ is m, then the ghost number of $|\chi\rangle$ is $m - 1$. Important special cases are states $|\psi\rangle$ with ghost number 0. If the ghost number is 0, then

$$U|\psi\rangle = 0$$

Looking at the form of U, namely, Eq. (6.7), we see that this implies that $b_k|\psi\rangle = 0$. Furthermore, a state annihilated by the ghost fields b_k cannot be annihilated by the ghost fields c^k. Since $u = \sum_i c^i b_i = \sum_i \delta^i_i - b_i c^i = n - \sum_i b_i c^i$ we see that

$$U|\psi\rangle = \left(n - \sum_i b_i c^i\right)|\psi\rangle = n|\psi\rangle - \sum_i b_i c^i|\psi\rangle$$

Hence if $b_k|\psi\rangle = 0$, the c^k cannot annihilate the state since this would result in a contradiction. When $b_k|\psi\rangle = 0$, it immediately follows that $U|\psi\rangle = 0$, but the above result shows that if $c^k|\psi\rangle = 0$ also, we would have $U|\psi\rangle = n|\psi\rangle \neq 0$, a contradiction.

The key result for states with zero ghost number is the following. Looking at Q in Eq. (6.5), it is clear that if $b_k|\psi\rangle = 0$ then

$$Q|\psi\rangle = \sum_i c^i K_i|\psi\rangle$$

This implies that for a state with zero ghost number

$$K_i |\psi\rangle = 0 \tag{6.9}$$

So, we can say that a state which is BRST invariant with zero ghost number is also invariant under the symmetry described by the generators K_i. Furthermore, if a state has ghost number zero, this tells us that the state is not a ghost state, hence we avoid negative probabilities.

BRST in String Theory-CFT

We will take a look at the BRST formalism in string theory by briefly considering two approaches. The derivation of this approach is based on the use of path integrals, which we are purposely avoiding due to the level of this text. So some results will simply be stated, the reader who is interested in their derivation is encouraged to check the references at the back of the book.

The application of BRST quantization to string theory can be done easily using conformal field theory. The advantage of this approach is that the critical dimension $D = 26$ arises in a straightforward manner. We work in the conformal gauge where we take $h_{\alpha\beta} = \eta_{\alpha\beta}$. In this case the energy-momentum tensor has a holomorphic component $T_{zz}(z)$ and an antiholomorphic component $T_{\bar{z}\bar{z}}(\bar{z})$ where $T_{zz}(z)$ was given in Eq. (5.33) as

$$T_{zz}(z) = \sum_{m=-\infty}^{\infty} \frac{L_m}{z^{m+2}}$$

In Example 5.5 we worked out the OPE of $T_{zz}(z)T_{ww}(w)$ and found

$$T_{zz}(z)T_{ww}(w) = \frac{D/2}{(z-w)^4} - \frac{2T_{ww}(w)}{(z-w)^2} - \frac{\partial_w T_{ww}(w)}{z-w}$$

Where for simplicity of notation we have omitted the multiplying ℓ_s^2 factor. The ghost fields are introduced as functions of a complex variable z as follows. We define

$$b(z)c(w) = \frac{1}{z-w} \qquad \bar{b}(\bar{z})\bar{c}(\bar{w}) = \frac{1}{\bar{z}-\bar{w}}$$

Next we write down an energy-momentum tensor $T_{gh}(z)$ for the ghost fields. This is given by

$$T_{gh}(z) = -2b(z)\partial_z c(z) - \partial_z b(z)c(z)$$

The conformal dimension [Eq. (6.4)] of the ghost fields follows from this definition. Using the ghost energy-momentum tensor together with the energy-momentum tensor for the string we can arrive at the BRST current:

$$j(z) = c(z)\left(T_{zz}(z) + \frac{1}{2} T_{gh}(z) \right) = c(z)T_{zz}(z) + c(z)\partial_z c(z)b(z)$$

The BRST charge is given by

$$Q = \int \frac{dz}{2\pi i} j(z)$$

Now, the central charge (i.e., the critical space-time dimension) comes from the leading term in the OPE of the energy-momentum tensor, which is

$$\frac{D/2}{(z-w)^4}$$

The presence of this extra term is called the *conformal anomaly* since it prevents the algebra from closing. So we would like to get rid of it. This is done by considering a total energy-momentum tensor, which is the sum of the string energy-momentum tensor and the ghost energy-momentum tensor, that is, $T = T_{zz}(z) + T_{gh}(z)$. It can be shown that the OPE of the ghost energy-momentum tensor is

$$T_{gh}(z)T_{gh}(w) = \frac{-13}{(z-w)^4} - \frac{2T_{gh}(w)}{(z-w)^2} - \frac{\partial_w T_{gh}(w)}{z-w}$$

Taking the leading term in this expression to be of the form $-(D/2)/(z-w)^4$, we see that the ghost fields contribute a central charge of -26 which precisely cancels the conformal anomaly that arises from the matter energy-momentum tensor. This result actually follows from the nilpotency requirement (i.e., $Q^2 = 0$) of the BRST charge.

BRST Transformations

Next we look at BRST quantization by considering a set of *BRST transformations* which are derived using a path integral approach. This is a bit beyond the level of the discussion used in the book, so we simply state the results. Working in light-cone coordinates, we define a ghost field c and an antighost field b where c has components c^+, c^-, and b has components b_{++} and b_{--}. We also introduce an energy-momentum tensor for the ghost fields $T_{\pm\pm}^{gh}$ with components given by

$$T_{++}^{gh} = i(2b_{++}\partial_+c^+ + \partial_+b_{++}c^+)$$
$$T_{--}^{gh} = i(2b_{--}\partial_-c^- + \partial_-b_{--}c^-)$$

The BRST transformations, using a small anticommuting operator ε are

$$\delta X^\mu = i\varepsilon(c^+\partial_+ + c^-\partial_-)X^\mu$$
$$\delta c^\pm = \pm i\varepsilon(c^+\partial_+ + c^-\partial_-)c^\pm$$
$$\delta b_{\pm\pm} = \pm i\varepsilon(T_{\pm\pm} + T_{\pm\pm}^{gh})$$

The action for the ghost fields is

$$S_{gh} = \int d^2\sigma(b_{++}\partial_-c^+ + b_{--}\partial_+c^-)$$

From which the following equations of motion follow:

$$\partial_-b_{++} = \partial_+b_{--} = \partial_-c^+ = \partial_+c^- = 0$$

We can write down modal expansions of the ghost fields. These are given by

$$c^+ = \sum_n \bar{c}_n e^{-in(\tau+\sigma)} \qquad c^- = \sum_n c_n e^{-in(\tau-\sigma)}$$
$$b_{++} = \sum_n \bar{b}_n e^{-in(\tau+\sigma)} \qquad b_{--} = \sum_n b_n e^{-in(\tau-\sigma)}$$

The modes satisfy the following anticommutation relation:

$$\{b_m, c_n\} = \delta_{m+n,0}$$

with $\{b_m, b_n\} = \{c_m, c_n\} = 0$. Virasoro operators are defined for the ghost fields using the modes. Using normal-ordered expansions, these are

$$L_m^{gh} = \sum_n (m-n) : b_{m+n}c_{-n} : \qquad \bar{L}_m^{gh} = \sum_n (m-n) : \bar{b}_{m+n}\bar{c}_{-n} :$$

To write down the total Virasoro operator for the "real" fields + ghost fields, we form a sum of the respective operators. That is

$$L_m^{tot} = L_m + L_m^{gh} - a\delta_{m,0}$$

where the last term on the right is the normal-ordering constant for $m = 0$. It can be shown that the commutation relation for the total Virasoro operator is of the form:

$$\left[L_m^{tot}, L_n^{tot}\right] = (m - n)L_{m+n}^{tot} + A(m)\delta_{m+n,0}$$

Notice that the presence of the term $A(m)$ on the right keeps us from obtaining a relation that preserves the classical Virasoro algebra. As such, this term is called an *anomaly*. The anomaly is determined in terms of two unknown constants which you might guess by now are D and a. It has the form

$$A(m) = \frac{D}{12}m(m^2 - 1) + \frac{1}{6}(m - 13m^3) + 2am$$

To make the anomaly vanish, we take $D = 26, a = 1$, which is consistent with the other results obtained so far in the book for bosonic string theory.

The BRST current is given by

$$j = cT + \frac{1}{2}:cT^{gh}: + \frac{3}{2}\partial^2 c$$

The BRST charge is given by the mode expansion:

$$Q = \sum_n c_n L_{-n} + \frac{1}{2}\sum_{m,n}(m - n):c_m c_n b_{-m-n}: -c_0$$

Using tedious algebra one can show that

$$Q^2 = \frac{1}{2}\sum\left(\left[L_m^{tot}, L_n^{tot}\right] - (m - n)L_{m+n}^{tot}\right)c_{-m}c_{-n} \approx \frac{1}{12}(D - 26)$$

Hence the requirement that $Q^2 = 0$ forces us to take $D = 26$.

Going back to the original BRST approach outlined in Eqs. (6.1) to (6.5), using the classical algebra for the Virasoro operators:

$$[L_m, L_n] = (m - n)L_{m+n}$$

We can identify the structure constants as

$$f_{mn}^k = (m-n)\delta_{m+n,k}$$

To see how the physical spectrum can be constructed in string theory, we consider the open string case. The states are built up from the ghost vacuum state. Let's call the ghost vacuum state $|\chi\rangle$. This state is annihilated by all positive ghost modes. Let $n > 0$, then

$$b_n|\chi\rangle = c_n|\chi\rangle = 0$$

The zero modes of the ghost fields are a special case. They can be used to build the physical states of the theory. Using the anticommutation relations [Eq. (6.3)], the zero modes satisfy

$$\{b_0, c_0\} = 1$$

Using Eq. (6.3) it should also be obvious that $b_0^2 = c_0^2 = 0$. We also require that $b_0|\psi\rangle = 0$ for physical states $|\psi\rangle$. Now we can construct a two-state system from the zero modes of the ghost states. The basis states are denoted by $|\uparrow\rangle, |\downarrow\rangle$. The ghost states act as

$$b_0|\downarrow\rangle = c_0|\uparrow\rangle = 0$$
$$b_0|\uparrow\rangle = |\downarrow\rangle \qquad c_0|\downarrow\rangle = |\uparrow\rangle$$

We choose $|\downarrow\rangle$ as the ghost vacuum state. To get the total state of the system, we take the tensor product of this state with the momentum state $|k\rangle$ to give $|\downarrow, k\rangle$. To generate a physical state, we act on it with the BRST charge Q. It can be shown that

$$Q|\downarrow, k\rangle = (L_0 - 1)c_0|\downarrow, k\rangle$$

The requirement that $Q|\downarrow, k\rangle = 0$ gives the mass-shell condition $L_0 - 1 = 0$, which describes the same Tachyon state we found in Chap. 4. Higher states can be generated. We will have mode operators for each of the three fields: the $X^\mu(\sigma, \tau)$ plus the two ghost fields. To get the first excited state, we act with α_{-1}, c_{-1}, and b_{-1} as follows:

$$|\psi\rangle = \left(\varsigma \cdot \alpha_{-1} + \xi_1 c_{-1} + \xi_2 b_{-1}\right)|\downarrow, k\rangle$$

The ξ_1 and ξ_2 are constants, but ς_μ is a vector with 26 components. In Chap. 4, we found that the first excited state was massless, so we expect the state to have physical degrees of freedom in the transverse directions. That is, it should have 24 independent components. To get rid of the extra parameters, we create a physical state with the requirement that $Q|\psi\rangle = 0$. It can be shown that

$$Q|\psi\rangle = 2\left(p^2 c_0 + (p \cdot \varsigma)c_{-1} + \xi_2 p \cdot \alpha_{-1}\right)|\downarrow, k\rangle$$

The requirement that $Q|\psi\rangle = 0$ enforces constraints on the parameters. With this general prescription, there are 26 positive norm states and 2 negative norm states.

We can eliminate the negative norm states by introducing some constraints. The first constraint is to take $p \cdot \varsigma = 0$ and $\xi_1 = 0$. This rids the theory of the negative norm states. Also note that $p^2 = 0$, which tells us that this is a massless state. We also have two zero-norm states:

$$k_\mu \alpha_{-1}^\mu |\downarrow, k\rangle \qquad \text{and} \qquad c_{-1}|\downarrow, k\rangle$$

These states are orthogonal to the physical states. Eliminating them gives us a state with 24 degrees of freedom, as expected for a massless state in 26 space-time dimensions.

No-Ghost Theorem

The *no-ghost theorem* is simply a statement of the results we have seen in Chap. 4 and here, namely, that if the number of space-time dimensions is given by $D = 26$, then negative norm states are eliminated from the theory.

Summary

In this chapter we introduced the BRST formalism and illustrated how it can be used to quantize strings. This is a more sophisticated approach than covariant quantization or light-cone quantization. It takes a middle ground, preserving manifest Lorentz invariance while living with ghost states. The approach makes the appearance of the critical $D = 26$ dimension simple to understand.

Quiz

1. Consider the Virasoro generators and calculate

$$\left[[L_i, L_j], L_k\right] + \left[[L_j, L_k], L_i\right] + \left[[L_k, L_i], L_j\right]$$

2. Suppose that the BRST current $j(z) = c(z)T_{zz}(z) + c(z)\partial_z c(z)b(z)$ is written as a normal ordered expression. Find $\frac{1}{2}\{Q, Q\}$.

3. Looking at your answer to question 2, how many scalar fields does the theory contain if we require that $Q^2 = 0$?

4. Use $L_0|\downarrow, k\rangle = 0$ to find k^2.

CHAPTER 7

RNS Superstrings

The real world as we know it described by the standard model of particle physics contains two general classes of particles. These are defined in terms of their spin angular momentum as follows. Those particles with whole integer spin are called *bosons,* while those with half-integer spin are called *fermions.* At the level of fundamental particles—electrons, neutrinos, quarks, photons, gauge bosons, and gluons—*matter particles* are fermions while *force-mediating particles* are bosons. A symmetry which relates fermions and bosons is called a *supersymmetry.*

The string theory we have described so far in the book consists only of bosons. Obviously, this cannot be a realistic theory that describes our universe since we see fermions in the everyday world. This tells us that the theory we have developed so far, starting with the Polyakov action, is not the whole story. The theory must be extended to include fermions. In addition, recall that when we quantized the theory, the ground state was a tachyon—a particle that travels faster than the speed of light. These states with negative mass squared are physically unrealistic. More importantly, a quantum theory with a tachyon has an unstable vacuum.

The remedy to this situation is to introduce *supersymmetry* into the theory. This will allow us to develop string theory so that fermions are included in our description of nature. We will also see that the unwanted tachyon state goes away. An interesting side effect of this effort will be that the critical dimension drops from 26 to 10. If you aren't familiar with the description of fermionic fields (and supersymmetry) in quantum theory, try reviewing your favorite quantum field theory book before tackling this chapter (*Quantum Field Theory Demystified* provides a relatively painless introduction).

The Superstring Action

We can proceed with a straightforward modification of the theory to include fermions using an approach called the *Ramond-Neveu-Schwarz (RNS) formalism*. This approach is supersymmetric on the worldsheet. Later we consider the *Green-Schwarz formalism,* which is supersymmetric in space-time. When the number of space-time dimensions is 10, these two approaches are equivalent.

The program we will follow can be done using basically the same which was applied in the bosonic case: introduce an action, find the equations of motion, and quantize the theory. However, this time we are going to include fermionic fields on the worldsheet. We start with the Polyakov action, first described in Eq. (2.27) and reproduced here in the conformal gauge:

$$S = -\frac{T}{2}\int d^2\sigma\, \partial_\alpha X^\mu\, \partial^\alpha X_\mu \tag{7.1}$$

To include free fermions in the theory using the RNS formalism, we add a kinetic energy term for a Dirac field to the lagrangian. That is, we include D free fermionic fields ψ^μ to the action, so that it assumes the form

$$S = -\frac{T}{2}\int d^2\sigma\left(\partial_\alpha X^\mu\, \partial^\alpha X_\mu - i\bar{\psi}^\mu \rho^\alpha \partial_\alpha \psi_\mu\right) \tag{7.2}$$

Again, if you are not familiar with Dirac fields, consult *Quantum Field Theory Demystified* or your own favorite quantum field theory text. The ρ^α are *Dirac matrices* on the worldsheet. Since the worldsheet has $1 + 1$ dimensions, the ρ^α are Dirac matrices in $1 + 1$ dimensions. Hence there are two such 2×2 matrices, which can be written in the form

$$\rho^0 = \begin{pmatrix} 0 & -i \\ i & 0 \end{pmatrix} \qquad \rho^1 = \begin{pmatrix} 0 & i \\ i & 0 \end{pmatrix} \tag{7.3}$$

using an appropriate choice of basis.

EXAMPLE 7.1

Show that the Dirac matrices on the worldsheet obey an anticommutation relation known as the *Dirac algebra*

$$\{\rho^\alpha, \rho^\beta\} = -2\eta^{\alpha\beta} \tag{7.4}$$

by explicit computation.

SOLUTION

This result is easy to verify. Since

$$\eta^{\alpha\beta} = \begin{pmatrix} -1 & 0 \\ 0 & 1 \end{pmatrix}$$

The Dirac algebra will be satisfied by the ρ^α if the following relations hold:

$$\{\rho^0, \rho^0\} = \rho^0\rho^0 + \rho^0\rho^0 = 2\rho^0\rho^0 = -2\eta^{00} = 2I$$
$$\{\rho^1, \rho^1\} = \rho^1\rho^1 + \rho^1\rho^1 = 2\rho^1\rho^1 = -2\eta^{11} = -2I$$
$$\{\rho^0, \rho^1\} = \rho^0\rho^1 + \rho^1\rho^0 = -2\eta^{01} = 0$$

and likewise for $\{\rho^1, \rho^0\}$. Now,

$$\rho^0\rho^0 = \begin{pmatrix} 0 & -i \\ i & 0 \end{pmatrix}\begin{pmatrix} 0 & -i \\ i & 0 \end{pmatrix} = \begin{pmatrix} 0\cdot 0 + (-i)\cdot i & 0\cdot(-i)+(-i)\cdot 0 \\ i\cdot 0 + 0\cdot i & i\cdot(-i)+0\cdot 0 \end{pmatrix} = \begin{pmatrix} 1 & 0 \\ 0 & 1 \end{pmatrix} = I$$

Hence the first relation $\{\rho^0, \rho^0\} = 2I$ is satisfied. We verify that the second relation $\{\rho^1, \rho^1\} = -2I$ is also satisfied:

$$\rho^1\rho^1 = \begin{pmatrix} 0 & i \\ i & 0 \end{pmatrix}\begin{pmatrix} 0 & i \\ i & 0 \end{pmatrix} = \begin{pmatrix} 0\cdot 0 + i\cdot i & 0\cdot i + i\cdot 0 \\ i\cdot 0 + 0\cdot i & i\cdot i + 0\cdot 0 \end{pmatrix} = \begin{pmatrix} -1 & 0 \\ 0 & -1 \end{pmatrix} = -I$$

Finally, noting that

$$\rho^0\rho^1 = \begin{pmatrix} 0 & -i \\ i & 0 \end{pmatrix}\begin{pmatrix} 0 & i \\ i & 0 \end{pmatrix} = \begin{pmatrix} 0\cdot 0 + (-i)\cdot i & 0\cdot i + (-i)\cdot 0 \\ i\cdot 0 + 0\cdot i & i\cdot i + 0\cdot 0 \end{pmatrix} = \begin{pmatrix} 1 & 0 \\ 0 & -1 \end{pmatrix}$$

$$\rho^1\rho^0 = \begin{pmatrix} 0 & i \\ i & 0 \end{pmatrix}\begin{pmatrix} 0 & -i \\ i & 0 \end{pmatrix} = \begin{pmatrix} 0\cdot 0 + i\cdot i & 0\cdot(-i)+i\cdot 0 \\ i\cdot 0 + 0\cdot i & i\cdot(-i)+0\cdot 0 \end{pmatrix} = \begin{pmatrix} -1 & 0 \\ 0 & 1 \end{pmatrix}$$

we see that $\{\rho^0, \rho^1\} = \{\rho^1, \rho^0\} = 0$.

MAJORANA SPINORS

The fields introduced in the action, $\psi^\mu = \psi^\mu(\sigma, \tau)$, are *two-component Majorana spinors* on the worldsheet. Given that they have two components, they are sometimes written with two indices ψ_A^μ, where $\mu = 0, 1, \ldots, D-1$ is the space-time index and $A = \pm$ is the spinor index. We can write ψ_A^μ as a column vector in the following way (suppressing the space-time index):

$$\psi = \begin{pmatrix} \psi_- \\ \psi_+ \end{pmatrix}$$

Under Lorentz transformations, these fields transform as vectors in space-time [recall that a contravariant vector field $V^\mu(x)$ is one that transforms as $V^\mu(x) \to V'^\mu(x') = \Lambda^\mu_{\ \nu} V^\nu(x)$ under $x^\mu \to x'^\mu = \Lambda^\mu_{\ \nu} x^\nu$ where $\Lambda^\mu_{\ \nu}$ is a Lorentz transformation].

Following the convention used with Dirac spinors in quantum field theory, we have the definition:

$$\bar{\psi}^\mu = (\psi^\dagger)^\mu \rho^0$$

Note that the definitions used here depend on the basis used to write down the Dirac matrices Eq. (7.3), and that other conventions are possible. We can also introduce a third Dirac matrix analogous to the γ_5 matrix you're familiar with from studies of the Dirac equation, which in this context we denote by ρ^3:

$$\rho^3 = \rho^0 \rho^1 = \begin{pmatrix} 1 & 0 \\ 0 & -1 \end{pmatrix}$$

It will be of interest to make left movers and right movers manifest. This can be done by recalling the following definitions:

$$\sigma^\pm = \tau \pm \sigma \tag{7.5}$$

$$\partial_\pm = \frac{1}{2}(\partial_\tau \pm \partial_\sigma) \tag{7.6}$$

$$\partial_\tau = \partial_+ + \partial_- \qquad \partial_\sigma = \partial_+ - \partial_- \tag{7.7}$$

EXAMPLE 7.2

Show that $\bar{\psi}^\mu \rho^\alpha \partial_\alpha \psi_\mu = 2(\psi_- \cdot \partial_+ \psi_- + \psi_+ \cdot \partial_- \psi_+)$.

SOLUTION

We can rewrite $\bar{\psi}^{\mu}\rho^{\alpha}\partial_{\alpha}\psi_{\mu}$ in a more enlightening way by expanding out the sum explicitly:

$$\bar{\psi}^{\mu}\rho^{\alpha}\partial_{\alpha}\psi_{\mu} = \bar{\psi}^{\mu}(\rho^{0}\partial_{0} + \rho^{1}\partial_{1})\psi_{\mu}$$

Now

$$\rho^{0}\partial_{0} = \begin{pmatrix} 0 & -i \\ i & 0 \end{pmatrix}\partial_{\tau}$$

$$\rho^{1}\partial_{1} = \begin{pmatrix} 0 & i \\ i & 0 \end{pmatrix}\partial_{\sigma}$$

So, the summation is

$$\rho^{0}\partial_{0} + \rho^{1}\partial_{1} = \begin{pmatrix} 0 & -i \\ i & 0 \end{pmatrix}\partial_{\tau} + \begin{pmatrix} 0 & i \\ i & 0 \end{pmatrix}\partial_{\sigma}$$

$$= \begin{pmatrix} 0 & -i(\partial_{\tau} - \partial_{\sigma}) \\ i(\partial_{\tau} + \partial_{\sigma}) & 0 \end{pmatrix} = \begin{pmatrix} 0 & -2i\partial_{-} \\ 2i\partial_{+} & 0 \end{pmatrix}$$

Hence,

$$\bar{\psi}^{\mu}\rho^{\alpha}\partial_{\alpha}\psi_{\mu} = \bar{\psi}^{\mu}(\rho^{0}\partial_{0} + \rho^{1}\partial_{1})\psi_{\mu}$$

$$= \bar{\psi}^{\mu}\begin{pmatrix} 0 & -2i\partial_{-} \\ 2i\partial_{+} & 0 \end{pmatrix}\psi_{\mu}$$

Now, we write out the components of the spinors. For simplicity, we suppress the space-time index for a moment. First, note that

$$\begin{pmatrix} 0 & -2i\partial_{-} \\ 2i\partial_{+} & 0 \end{pmatrix}\psi = \begin{pmatrix} 0 & -2i\partial_{-} \\ 2i\partial_{+} & 0 \end{pmatrix}\begin{pmatrix} \psi_{-} \\ \psi_{+} \end{pmatrix} = 2i\begin{pmatrix} -\partial_{-}\psi_{+} \\ \partial_{+}\psi_{-} \end{pmatrix}$$

Using $\bar{\psi}^{\mu} = (\psi^{\dagger})^{\mu}\rho^{0}$, we have

$$\bar{\psi}^{\mu}\begin{pmatrix} 0 & -2i\partial_{-} \\ 2i\partial_{+} & 0 \end{pmatrix}\psi_{\mu} = \begin{pmatrix} \psi_{-}^{\mu} & \psi_{+}^{\mu} \end{pmatrix}\begin{pmatrix} 0 & -i \\ i & 0 \end{pmatrix}2i\begin{pmatrix} -\partial_{-}\psi_{+\mu} \\ \partial_{+}\psi_{-\mu} \end{pmatrix}$$

$$= \begin{pmatrix} \psi_{-}^{\mu} & \psi_{+}^{\mu} \end{pmatrix}\begin{pmatrix} 2\partial_{+}\psi_{-\mu} \\ 2\partial_{-}\psi_{+\mu} \end{pmatrix} = 2\psi_{-}^{\mu}\partial_{+}\psi_{-\mu} + 2\psi_{+}^{\mu}\partial_{-}\psi_{+\mu}$$

$$= 2(\psi_{-}\cdot\partial_{+}\psi_{-} + \psi_{+}\cdot\partial_{-}\psi_{+})$$

The result obtained in Example 7.2 allows us to write the fermionic part of the action in a relatively simple way. Denoting the fermionic action by S_F we have

$$
\begin{aligned}
S_F &= -\frac{T}{2}\int d^2\sigma(-i\bar{\psi}^\mu\rho^\alpha\partial_\alpha\psi_\mu)\\
&= -\frac{T}{2}\int d^2\sigma(-2i)(\psi_-\cdot\partial_+\psi_- + \psi_+\cdot\partial_-\psi_+)\\
&= iT\int d^2\sigma(\psi_-\cdot\partial_+\psi_- + \psi_+\cdot\partial_-\psi_+)
\end{aligned}
$$

It can be shown that by varying the fermionic action S_F, one can obtain the free field Dirac equations of motion:

$$
\partial_+\psi_-^\mu = \partial_-\psi_+^\mu = 0 \tag{7.8}
$$

The Majorana field ψ_-^μ describes right movers while the Majorana field ψ_+^μ describes left movers.

SUPERSYMMETRY TRANSFORMATIONS ON THE WORLDSHEET

Now, we introduce a *supersymmetry* (SUSY for short) *transformation parameter* which is denoted by ε. This infinitesimal object is also a Majorana spinor, which has real, constant components given by

$$
\varepsilon = \begin{pmatrix} \varepsilon_- \\ \varepsilon_+ \end{pmatrix}
$$

Since the components of ε are taken to be constant, this represents a *global symmetry* of the worldsheet. If it were a local symmetry, it would depend on the coordinates (σ, τ). Furthermore, the components of ε are *Grassman numbers*. Two Grassmann numbers a, b anticommute such that $ab + ba = 0$.

Now we use ε to define our symmetry. The action which includes the fermionic fields is invariant under the supersymmetry transformations:

$$
\begin{aligned}
\delta X^\mu &= \bar{\varepsilon}\psi^\mu \\
\delta\psi^\mu &= -i\rho^\alpha\partial_\alpha X^\mu\varepsilon
\end{aligned} \tag{7.9}
$$

Using $\delta\psi^\mu$, we also find that $\delta\bar{\psi}^\mu = \overline{-i\rho^\alpha\partial_\alpha X^\mu\varepsilon} = i\bar{\varepsilon}\rho^\alpha\partial_\alpha X^\mu$. Notice that this takes the free boson fields into fermionic fields, and vice versa. We can relate individual components as follows. First, we have

$$
\delta X^\mu = \bar{\varepsilon}\psi^\mu = (\varepsilon_- \quad \varepsilon_+)\begin{pmatrix} \psi_-^\mu \\ \psi_+^\mu \end{pmatrix} = \varepsilon_-\psi_-^\mu + \varepsilon_+\psi_+^\mu \tag{7.10}
$$

In the second case, we have

$$\delta\psi^{\mu} = \begin{pmatrix} \delta\psi^{\mu}_{-} \\ \delta\psi^{\mu}_{+} \end{pmatrix}$$

and

$$\rho^{\alpha}\partial_{\alpha}X^{\mu}\varepsilon = \rho^{0}\partial_{\tau}X^{\mu}\varepsilon + \rho^{1}\partial_{\sigma}X^{\mu}\varepsilon$$
$$= (\rho^{0}\partial_{\tau} + \rho^{1}\partial_{\sigma})X^{\mu}\varepsilon$$
$$= \left[\begin{pmatrix} 0 & -i \\ i & 0 \end{pmatrix}\partial_{\tau} + \begin{pmatrix} 0 & i \\ i & 0 \end{pmatrix}\partial_{\sigma}\right]X^{\mu}\varepsilon$$
$$= \begin{pmatrix} 0 & -i(\partial_{\tau} - \partial_{\sigma}) \\ i(\partial_{\tau} + \partial_{\sigma}) & 0 \end{pmatrix}X^{\mu}\varepsilon$$
$$= \begin{pmatrix} 0 & -2i\partial_{-} \\ 2i\partial_{+} & 0 \end{pmatrix}X^{\mu}\varepsilon$$

Hence,

$$\delta\psi^{\mu}_{-} = -2\partial_{-}X^{\mu}\varepsilon_{+} \qquad (7.11)$$
$$\delta\psi^{\mu}_{+} = 2\partial_{+}X^{\mu}\varepsilon_{-} \qquad (7.12)$$

Conserved Currents

At this point, we need to identify the conserved currents associated with the action in Eq. (7.2). Before tackling supersymmetry, let's review how to calculate a conserved current by looking at momentum. You can practice in the Chapter Quiz by looking at Lorentz transformations. Let's start with a simple example to remind ourselves of the method.

EXAMPLE 7.3
Consider the action $S = -T/2\int d^{2}\sigma\,(\partial_{\alpha}X^{\mu}\partial^{\alpha}X_{\mu} - i\overline{\psi}^{\mu}\rho^{\alpha}\partial_{\alpha}\psi_{\mu})$ and find the conserved current associated with translational invariance.

SOLUTION

Let's write down the lagrangian, which is

$$L = -\frac{T}{2}(\partial_\alpha X^\mu \partial^\alpha X_\mu - i\bar{\psi}^\mu \rho^\alpha \partial_\alpha \psi_\mu)$$

To examine translational invariance, we let $X^\mu \to X^\mu + a^\mu$ where a^μ is an *infinitesimal* parameter. A key insight into the fact that a^μ is infinitesimal is that we can drop terms that are second order in a^μ. Taking $X^\mu \to X^\mu + a^\mu$ changes the lagrangian as follows:

$$L \to -\frac{T}{2}\Big[\partial_\alpha(X^\mu + a^\mu)\partial^\alpha(X_\mu + a_\mu) - i\bar{\psi}^\mu \rho^\alpha \partial_\alpha \psi_\mu\Big]$$

$$= -\frac{T}{2}\Big[(\partial_\alpha X^\mu + \partial_\alpha a^\mu)(\partial^\alpha X_\mu + \partial^\alpha a_\mu)\Big] + L_F$$

$$= -\frac{T}{2}\Big[\partial_\alpha X^\mu \partial^\alpha X_\mu + \partial_\alpha X^\mu \partial^\alpha a_\mu + \partial_\alpha a^\mu \partial^\alpha X_\mu + \partial_\alpha a^\mu \partial^\alpha a_\mu\Big] + L_F$$

$$= -\frac{T}{2}\Big[\partial_\alpha X^\mu \partial^\alpha X_\mu + \partial_\alpha X^\mu \partial^\alpha a_\mu + \partial_\alpha a^\mu \partial^\alpha X_\mu\Big] + L_F$$

(drop second-order term $\partial_\alpha a^\mu \partial^\alpha a_\mu$)

$$= L - \frac{T}{2}\Big[\partial_\alpha X^\mu \partial^\alpha a_\mu + \partial_\alpha a^\mu \partial^\alpha X_\mu\Big]$$

(add in $\partial_\alpha X^\mu \partial^\alpha X_\mu$ to L_F to get total lagrangian)

Note that the term $i\bar{\psi}^\mu \rho^\alpha \partial_\alpha \psi_\mu \propto L_F$ is unaffected by $X^\mu \to X^\mu + a^\mu$. Now, we focus on the leftover extra term:

$$\delta L = \frac{T}{2}\Big[\partial_\alpha X^\mu \partial^\alpha a_\mu + \partial_\alpha a^\mu \partial^\alpha X_\mu\Big]$$

We will manipulate this expression to get our conserved current, which will be a term multiplying $\partial^\alpha a_\mu$. We can fix up this expression doing some index gymnastics, which is a good exercise for us to go through given the level of this book. For more practice doing this, consult *Relativity Demystified*.

We want to fix up the second term so that it looks like the first term. We do this by raising and lowering indices with the metric and using the fact that

$$\eta^{\mu\nu}\eta_{\nu\lambda} = \delta^\mu_\lambda \qquad h^{\alpha\beta}h_{\beta\gamma} = \delta^\alpha_\gamma$$

Now, the first thing to notice is that the order of the derivatives doesn't matter. So, the first step is to write

$$\partial_\alpha a^\mu \partial^\alpha X_\mu = \partial^\alpha X_\mu \partial_\alpha a^\mu$$

Next, let's raise the space-time index on X_μ and lower it on a^μ. Since these are space-time indices, we use the Minkowski metric to do this:

$$\partial^\alpha X_\mu \partial_\alpha a^\mu = \partial^\alpha (\eta_{\mu\nu} X^\nu) \partial_\alpha (\eta^{\mu\lambda} a_\lambda)$$

The metric is not space-time dependent (in the flat space or Minkowski space-time we are considering here), so we can pull the metric terms outside of the derivatives. You might recognize that this is actually true in general—because the derivatives are with respect to worldsheet coordinates, but the metric, if it depends on coordinates, is space-time dependent. So,

$$\partial^\alpha (\eta_{\mu\nu} X^\nu) \partial_\alpha (\eta^{\mu\lambda} a_\lambda) = \eta_{\mu\nu} \eta^{\mu\lambda} \partial^\alpha X^\nu \partial_\alpha a_\lambda$$
$$= \delta_\nu^\lambda \partial^\alpha X^\nu \partial_\alpha a_\lambda = \partial^\alpha X^\nu \partial_\alpha a_\nu$$

The index ν is a repeated or dummy index, so we can call it what we like. Let's change it to match the first term in $\delta L = T/2 [\partial_\alpha X^\mu \partial^\alpha a_\mu + \partial_\alpha a^\mu \partial^\alpha X_\mu]$:

$$\partial^\alpha X^\nu \partial_\alpha a_\nu = \partial^\alpha X^\mu \partial_\alpha a_\mu$$

Now, we repeat the process for the indices on the derivatives. This time, the indices are worldsheet indices. So, we obtain

$$\partial^\alpha X^\mu \partial_\alpha a_\mu = h^{\alpha\beta} \partial_\beta X^\mu h_{\alpha\gamma} \partial^\gamma a_\mu$$
$$= h^{\alpha\beta} h_{\alpha\gamma} \partial_\beta X^\mu \partial^\gamma a_\mu$$
$$= \delta_\gamma^\beta \partial_\beta X^\mu \partial^\gamma a_\mu$$
$$= \partial_\beta X^\mu \partial^\beta a_\mu = \partial_\alpha X^\mu \partial^\alpha a_\mu$$

And so, the variation in the lagrangian reduces to

$$\delta L = \frac{T}{2} (\partial_\alpha X^\mu \partial^\alpha a_\mu + \partial_\alpha a^\mu \partial^\alpha X_\mu) = T \partial_\alpha X^\mu \partial^\alpha a_\mu$$

The leftover term multiplying the infinitesimal $\partial^\alpha a_\mu$ is our conserved current. Being that we started with a translation of space-time coordinates, we identify this as the momentum:

$$P_\alpha^\mu = T\partial_\alpha X^\mu$$

With Example 7.3 in mind, we can easily find the conserved supercurrent, which is the conserved current associated with the supersymmetry transformation. Let's just grind it out. Starting with $L = -T/2(\partial_\alpha X^\mu \partial^\alpha X_\mu - i\bar\psi^\mu \rho^\alpha \partial_\alpha \psi_\mu)$, we have

$$\delta L = -\frac{T}{2}\Big[2\partial_\alpha(\delta X^\mu)\partial^\alpha X_\mu - i(\delta\bar\psi^\mu)\rho^\alpha\partial_\alpha\psi_\mu - i\bar\psi^\mu\rho^\alpha\partial_\alpha(\delta\psi_\mu)\Big]$$

$$= -\frac{T}{2}\Big[2\partial_\alpha(\bar\varepsilon\psi^\mu)\partial^\alpha X_\mu + (\bar\varepsilon\rho^\beta\partial_\beta X^\mu)\rho^\alpha\partial_\alpha\psi_\mu - \bar\psi^\mu\rho^\alpha\partial_\alpha(\rho^\beta\partial_\beta X^\mu\varepsilon)\Big]$$

$$= -\frac{T}{2}\Big[2\partial_\alpha(\bar\varepsilon\psi^\mu)\partial^\alpha X_\mu - \partial_\alpha(\bar\varepsilon\rho^\beta\partial_\beta X^\mu)\rho^\alpha\psi_\mu - \bar\psi^\mu\rho^\alpha\partial_\alpha(\rho^\beta\partial_\beta X_\mu\varepsilon)\Big]$$

$$= -T\Big[\partial_\alpha(\bar\varepsilon\psi^\mu)\partial^\alpha X_\mu - \partial_\alpha(\bar\varepsilon\rho^\beta\partial_\beta X^\mu)\rho^\alpha\psi_\mu\Big]$$

$$= -T\Big[\partial_\alpha(\bar\varepsilon\psi^\mu)\partial^\alpha X_\mu - \partial_\alpha\bar\varepsilon(\rho^\beta\partial_\beta X^\mu)\rho^\alpha\psi_\mu - \bar\varepsilon\rho^\beta\rho^\alpha(\partial_\alpha\partial_\beta X^\mu)\psi_\mu\Big]$$

$$= -T\Big[\partial_\alpha(\bar\varepsilon\psi^\mu\partial^\alpha X_\mu) - \bar\varepsilon\psi^\mu\partial_\alpha\partial^\alpha X_\mu - \partial_\alpha\bar\varepsilon(\rho^\beta\rho^\alpha\psi^\mu\partial_\beta X_\mu) + \bar\varepsilon(\partial_\alpha\partial^\alpha X_\mu)\psi^\mu\Big]$$

$$= -T\Big[\partial_\alpha(\bar\varepsilon\psi^\mu\partial^\alpha X_\mu) - \partial_\alpha\bar\varepsilon(\rho^\beta\rho^\alpha\psi^\mu\partial_\beta X_\mu)\Big]$$

The first term is a total derivative, so it does not contribute to the variation of the action. So we identify the conserved current with the second term. It is taken to be

$$J_\alpha^\mu = \frac{1}{2}\rho^\beta\rho_\alpha\psi^\mu\partial_\beta X_\mu \tag{7.13}$$

The Energy-Momentum Tensor

The next item of interest in our description of strings with worldsheet supersymmetry is the derivation of the energy-momentum tensor. The energy-momentum tensor is associated with translation symmetry on the worldsheet. Consider an infinitesimal translation ε^α which is used to vary the worldsheet coordinates as

$$\sigma^\alpha \to \sigma^\alpha + \varepsilon^\alpha$$

We can write the change of the bosonic fields X^μ by basically writing down their Taylor expansion:

$$X^\mu \to X^\mu + \varepsilon^\alpha \partial_\alpha X^\mu \qquad (7.14)$$

A similar relation holds for the fermionic fields:

$$\psi^\mu \to \psi^\mu + \varepsilon^\alpha \partial_\alpha \psi^\mu \qquad (7.15)$$

With this in mind, we again follow the Noether procedure. Vary the action as if ε^α depended on the worldsheet coordinates, and look for terms multiplied by $\partial_\beta \varepsilon^\alpha$. At the end we consider ε^α to be constant so that term vanishes from the action—the term which multiplies $\partial_\beta \varepsilon^\alpha$ will be the energy-momentum tensor that we seek.

We proceed in two parts. Let's take a look at the fermionic part of the lagrangian first. We have

$$L_F = -\frac{i}{2} \bar\psi^\mu \rho^\alpha \partial_\alpha \psi_\mu$$

Using Eq. (7.15), we vary this term as follows:

$$\delta L_F = -\frac{i}{2} (\delta\bar\psi^\mu) \rho^\alpha \partial_\alpha \psi_\mu - \frac{i}{2} \bar\psi^\mu \rho^\alpha \partial_\alpha (\delta\psi_\mu)$$

$$= -\frac{i}{2} (\varepsilon^\beta \partial_\beta \bar\psi^\mu) \rho^\alpha \partial_\alpha \psi_\mu - \frac{i}{2} \bar\psi^\mu \rho^\alpha \partial_\alpha (\varepsilon^\beta \partial_\beta \psi_\mu)$$

Let's apply the product rule and carry out the derivative on the second term:

$$-\frac{i}{2} (\varepsilon^\beta \partial_\beta \bar\psi^\mu) \rho^\alpha \partial_\alpha \psi_\mu - \frac{i}{2} \bar\psi^\mu \rho^\alpha \partial_\alpha (\varepsilon^\beta \partial_\beta \psi_\mu)$$

$$= -\frac{i}{2} (\varepsilon^\beta \partial_\beta \bar\psi^\mu) \rho^\alpha \partial_\alpha \psi_\mu - \frac{i}{2} \bar\psi^\mu \rho^\alpha \partial_\alpha \varepsilon^\beta \partial_\beta \psi_\mu - \frac{i}{2} \bar\psi^\mu \rho^\alpha \varepsilon^\beta \partial_\alpha \partial_\beta \psi_\mu$$

Now, the variation actually takes place as a variation of the action S, so we can integrate by parts. We do this on the last term to move one of the derivatives off $\partial_\alpha \partial_\beta \psi_\mu$. Integration by parts introduces a sign change, so we get

$$-\frac{i}{2} (\varepsilon^\beta \partial_\beta \bar\psi^\mu) \rho^\alpha \partial_\alpha \psi_\mu - \frac{i}{2} \bar\psi^\mu \rho^\alpha \partial_\alpha \varepsilon^\beta \partial_\beta \psi_\mu - \frac{i}{2} \bar\psi^\mu \rho^\alpha \varepsilon^\beta \partial_\alpha \partial_\beta \psi_\mu$$

$$= -\frac{i}{2} (\varepsilon^\beta \partial_\beta \bar\psi^\mu) \rho^\alpha \partial_\alpha \psi_\mu - \frac{i}{2} \bar\psi^\mu \rho^\alpha \partial_\alpha \varepsilon^\beta \partial_\beta \psi_\mu + \frac{i}{2} \partial_\beta (\bar\psi^\mu \rho^\alpha \varepsilon^\beta) \partial_\alpha \psi_\mu$$

$$= -\frac{i}{2} (\varepsilon^\beta \partial_\beta \bar\psi^\mu) \rho^\alpha \partial_\alpha \psi_\mu - \frac{i}{2} \bar\psi^\mu \rho^\alpha \partial_\alpha \varepsilon^\beta \partial_\beta \psi_\mu + \frac{i}{2} \varepsilon^\beta \partial_\beta \bar\psi^\mu \rho^\alpha \partial_\alpha \psi_\mu + \frac{i}{2} \bar\psi^\mu \rho^\alpha \partial_\beta \varepsilon^\beta \partial_\alpha \psi_\mu$$

The divergence term $\partial_\beta \varepsilon^\beta$ is not going to contribute anything, so we drop it. The first and third terms cancel, leaving us with

$$\delta L_F = \partial_\alpha \varepsilon^\beta \left(-\frac{i}{2} \bar{\psi}^\mu \rho^\alpha \partial_\beta \psi_\mu \right)$$

This is what we want, because terms that multiply $\partial_\alpha \varepsilon^\beta$ are going to be terms that make up the energy-momentum tensor $T_{\alpha\beta}$. This isn't quite right, because we want it to be symmetric. So, we take

$$\delta L_F = \partial_\alpha \varepsilon^\beta \left(-\frac{i}{4} \bar{\psi}^\mu \rho^\alpha \partial_\beta \psi_\mu - \frac{i}{4} \bar{\psi}^\mu \rho^\beta \partial_\alpha \psi_\mu \right) \tag{7.16}$$

In the Chapter Quiz, you will derive an expression for the bosonic part of the energy-momentum tensor. When all is said and done

$$T_{\alpha\beta} = \partial_\alpha X^\mu \partial_\beta X_\mu + \frac{i}{4} \bar{\psi}^\mu \rho_\alpha \partial_\beta \psi_\mu + \frac{i}{4} \bar{\psi}^\mu \rho_\beta \partial_\alpha \psi_\mu - (\text{Trace}) \tag{7.17}$$

The "Trace" is explicitly removed to ensure that $T_{\alpha\beta}$ remains traceless as required for scale invariance.

The energy-momentum tensor and supercurrent can be written compactly using worldsheet light-cone coordinates. The energy-momentum tensor has two nonzero components given by

$$T_{++} = \partial_+ X_\mu \partial_+ X^\mu + \frac{i}{2} \psi_+^\mu \partial_+ \psi_{+\mu} \qquad T_{--} = \partial_- X_\mu \partial_- X^\mu + \frac{i}{2} \psi_-^\mu \partial_- \psi_{-\mu} \tag{7.18}$$

The components of the supercurrent are

$$J_+ = \psi_+^\mu \partial_+ X_\mu \qquad J_- = \psi_-^\mu \partial_- X_\mu \tag{7.19}$$

The equations of motion for the fermion fields are

$$\partial_+ \psi_-^\mu = \partial_- \psi_+^\mu = 0 \tag{7.20}$$

Together with the equations of motion of the boson fields

$$\partial_+ \partial_- X^\mu = 0 \tag{7.21}$$

We obtain conservation laws for the energy-momentum tensor:

$$\partial_- T_{++} = \partial_+ T_{--} = 0 \tag{7.22}$$

EXAMPLE 7.4

Show that the equations of motion for the fermion and boson fields lead to conservation of the supercurrent.

SOLUTION

We start with J_+ and consider the derivative $\partial_- J_+$. We have:

$$
\begin{aligned}
\partial_- J_+ &= \partial_- (\psi_+^\mu \partial_+ X_\mu) \\
&= (\partial_- \psi_+^\mu) \partial_+ X_\mu + \psi_+^\mu (\partial_- \partial_+ X_\mu) \\
&= 0
\end{aligned}
$$

The result was readily obtained using Eqs. (7.20) and (7.21). Now taking J_-, we obtain a second conservation equation by calculating $\partial_+ J_-$ which gives

$$
\begin{aligned}
\partial_+ J_- &= \partial_+ (\psi_-^\mu \partial_- X_\mu) \\
&= \left(\partial_+ \psi_-^\mu\right) \partial_- X_\mu + \psi_-^\mu \left(\partial_+ \partial_- X_\mu\right) \\
&= \psi_-^\mu (\partial_- \partial_+ X_\mu) = 0
\end{aligned}
$$

EXAMPLE 7.5

Show that $\partial_- T_{++} = 0$.

SOLUTION

Using $T_{++} = \partial_+ X_\mu \partial_+ X^\mu + i/2 \psi_+^\mu \partial_+ \psi_{+\mu}$, we find

$$
\begin{aligned}
\partial_- T_{++} &= \partial_- \left(\partial_+ X_\mu \partial_+ X^\mu + \frac{i}{2} \psi_+^\mu \partial_+ \psi_{+\mu} \right) \\
&= (\partial_- \partial_+ X_\mu) \partial_+ X^\mu + \partial_+ X_\mu (\partial_- \partial_+ X^\mu) + \frac{i}{2} \left(\partial_- \psi_+^\mu\right) \partial_+ \psi_{+\mu} + \frac{i}{2} \psi_+^\mu \partial_- \partial_+ \psi_{+\mu} \\
&= \frac{i}{2} \psi_+^\mu \partial_- \partial_+ \psi_{+\mu} = \frac{i}{2} \psi_+^\mu \partial_+ \partial_- \psi_{+\mu} = 0
\end{aligned}
$$

To obtain this result, we applied Eqs. (7.20) and (7.21) together with the commutativity of partial derivatives.

Mode Expansions and Boundary Conditions

The final step in putting together the classical physics of the RNS superstring follows the program used in the bosonic case—we need to apply boundary conditions and write down the mode expansions. Specifically, we need to apply boundary conditions for the fermionic fields. It is simplest to continue working in light-cone coordinates and vary the fermionic part of the action. Before doing this, it can be helpful to review some elementary calculus.

Recall integration by parts:

$$\int_a^b f(x)\frac{dg}{dx}dx = fg\Big|_a^b - \int_a^b \frac{df}{dx}g(x)\,dx$$

The product fg is called the boundary term. When we vary the fermionic action, we are going to obtain boundary terms for the fields ψ_+, so we need to specify boundary conditions so that the variation in the action vanishes. The fermionic part of the action in light-cone coordinates, modulo a few constants and ignoring the space-time index is

$$S_F = \int d^2\sigma(\psi_-\partial_+\psi_- + \psi_+\partial_-\psi_+) \tag{7.23}$$

For simplicity, let's consider one piece of this expression and vary it. We obtain

$$\delta\int d^2\sigma\,\psi_+\partial_-\psi_+ = \int d^2\sigma[\delta\psi_+\partial_-\psi_+ + \psi_+\partial_-(\delta\psi_+)]$$

Following the usual procedure applied in field theory, we want to move the derivative off the $\delta\psi_+$ term. This can be done using integration by parts. When this is done, we pick up a boundary term:

$$\int d^2\sigma\,\psi_+\partial_-(\delta\psi_+) = \int_{-\infty}^{\infty} d\tau\,\psi_+\delta\psi_+\Big|_{\sigma=0}^{\sigma=\pi} - \int d^2\sigma\partial_-\psi_+\delta\psi_+$$

A similar expression arises from the variation of the other term. All together, the boundary terms obtained by varying the action are

$$\delta S_F = \int_{-\infty}^{\infty} d\tau\left\{(\psi_+\delta\psi_+ - \psi_-\delta\psi_-)\Big|_{\sigma=\pi} - (\psi_+\delta\psi_+ - \psi_-\delta\psi_-)\Big|_{\sigma=0}\right\} \tag{7.24}$$

OPEN STRING BOUNDARY CONDITIONS

When varying the action, the boundary terms must vanish in order to maintain Lorentz invariance. In the case of open string, the boundary terms $\sigma = 0$ and $\sigma = \pi$ must both vanish independently. We can obtain

$$\psi_+ \delta\psi_+ - \psi_- \delta\psi_- = 0$$

at $\sigma = 0$ if we take

$$\psi_\mu^+(0, \tau) = \psi_\mu^-(0, \tau) \tag{7.25}$$

Now in general, $\psi_+ = \pm\psi_-$ will make the boundary terms vanish, but typical convention is to fix the boundary condition at $\sigma = 0$ using Eq. (7.25). This leaves the choice of sign at $\sigma = \pi$ ambiguous. Depending on the sign we choose, we obtain two different boundary conditions. *Ramond* or R boundary conditions are given by the choice

$$\psi_\mu^+(\pi, \tau) = \psi_\mu^-(\pi, \tau) \qquad \text{(Ramond)} \tag{7.26}$$

The other choice we can make is known as *Neveau-Schwarz* or NS boundary conditions:

$$\psi_\mu^+(\pi, \tau) = -\psi_\mu^-(\pi, \tau) \qquad \text{(Neveau-Schwarz)} \tag{7.27}$$

We often refer to the boundary conditions chosen as the *sector*. The choice of boundary conditions has dramatic consequences. In particular

- The R sector gives rise to string states that are space-time fermions.
- The NS sector gives rise to string states that are space-time bosons.

OPEN STRING MODE EXPANSIONS

We consider the R sector first. The mode expansions are

$$\psi_-^\mu(\sigma, \tau) = \frac{1}{\sqrt{2}} \sum_n d_n^\mu e^{-in(\tau-\sigma)}$$

$$\psi_+^\mu(\sigma, \tau) = \frac{1}{\sqrt{2}} \sum_n d_n^\mu e^{-in(\tau+\sigma)} \tag{7.28}$$

The *Majorana condition* is the requirement that the fermionic fields are real. This forces us to take

$$d^\mu_{-n} = \left(d^\mu_n\right)^\dagger \tag{7.29}$$

Here, the summation index is an integer and so runs $n = 0, \pm 1, \pm 2, \ldots$.

The NS sector results in different mode expansions, as you might guess, since this gives rise to different string states. The expansions are

$$\psi^\mu_-(\sigma, \tau) = \frac{1}{\sqrt{2}} \sum_r b^\mu_r e^{-ir(\tau-\sigma)}$$
$$\psi^\mu_+(\sigma, \tau) = \frac{1}{\sqrt{2}} \sum_r b^\mu_r e^{-ir(\tau+\sigma)} \tag{7.30}$$

This is more than simple notational gymnastics. The summations in the NS sector are quite different than those for the R sector, because here we take

$$r = \pm\frac{1}{2}, \ \pm\frac{3}{2}, \ \pm\frac{5}{2}, \ \ldots \tag{7.31}$$

CLOSED STRING BOUNDARY CONDITIONS

In the case of closed strings, we can apply periodic or antiperiodic boundary conditions. These are given by

$$\psi_\pm(\sigma, \tau) = \psi_\pm(\sigma+\pi,\tau) \qquad \text{(periodic boundary condition)}$$
$$\psi_\pm(\sigma, \tau) = -\psi_\pm(\sigma+\pi,\tau) \qquad \text{(antiperiodic boundary condition)} \tag{7.32}$$

CLOSED STRING MODE EXPANSIONS

The boundary conditions in Eq. (7.32) can be applied separately to left and right movers. The mode expansions are

$$\psi^\mu_+(\sigma, \tau) = \sum_r \tilde{d}^\mu_r e^{-2ir(\tau+\sigma)}$$
$$\psi^\mu_-(\sigma, \tau) = \sum_r d^\mu_r e^{-2ir(\tau-\sigma)} \tag{7.33}$$

If we choose the R sector, then following the open string case

$$r = 0, \pm 1, \pm 2, \ldots \qquad (7.34)$$

On the other hand, if we choose the NS sector, then

$$r = \pm \frac{1}{2}, \pm \frac{3}{2}, \cdots \qquad (7.35)$$

We can choose either sector for left and right movers independently. If the sectors match for left and right movers, we obtain space-time bosons. If the sectors are different for left and right movers, then we obtain space-time fermions. That is,

- Choosing the NS sector for left movers and the NS sector for right movers gives space-time bosons.
- Choosing the R sector for left movers and the R sector for right movers gives space-time bosons.
- Choosing the NS sector for left movers and the R sector for right movers gives space-time fermions.
- Choosing the R sector for left movers and the NS sector for right movers gives space-time fermions.

Super-Virasoro Generators

When we quantize the theory we will need *super-Virasoro operators*. These are generalizations of what we have already worked out for bosonic string theory. We extend the idea in this case to include a fermionic operator. That is,

$$L_m \rightarrow L_m^{(B)} + L_m^{(F)}$$

It can be shown that the following definition will work:

$$L_m = \frac{1}{2} \sum_{n=-\infty}^{\infty} \alpha_{m-n} \cdot \alpha_n + \frac{1}{4} \sum_r (2r - m) b_{-r} b_{m+r} \qquad (7.36)$$

In addition, in superstring theory we have a second generator that arises from the supercurrent

$$G_r = \frac{\sqrt{2}}{\pi} \int_{-\pi}^{\pi} d\sigma\, e^{ir\sigma} J_+ = \sum_{m=-\infty}^{\infty} \alpha_m \cdot b_{r+m} \tag{7.37}$$

for the NS sector, while we take

$$F_m = \sum_n \alpha_{-n} \cdot d_{m+n} \tag{7.38}$$

for the R sector. Here, m and n are integers while $r = \pm 1/2, \pm 3/2, \ldots$.

Canonical Quantization

Now we are ready to quantize the theory, and canonical quantization is not so bad because fermions are simple to deal with. The condition on the modes for the bosonic string was the commutator:

$$\left[\alpha_m^\mu, \alpha_n^\nu\right] = m\delta_{m+n,0}\eta^{\mu\nu} \tag{7.39}$$

This relation is supplemented by a similar commutator for the $\bar{\alpha}$'s in the case of closed strings. For the supersymmetric theory, we need to supplement Eq. (7.39) with relations for the fermionic modes. You will recall from your studies of quantum field theory that fermionic fields satisfy anticommutation relations. In our case the Majorana fields will satisfy the equal time anticommutation relation:

$$\{\psi_A^\mu(\sigma,\tau), \psi_B^\nu(\sigma',\tau)\} = \pi\eta^{\mu\nu}\delta_{AB}\delta(\sigma-\sigma') \tag{7.40}$$

In terms of the modes, we will have the following sets of anticommutation relations depending on the sector used:

$$\begin{aligned}
\left\{b_r^\mu, b_s^\nu\right\} &= \eta^{\mu\nu}\delta_{r+s,0} \\
\left\{d_m^\mu, d_n^\nu\right\} &= \eta^{\mu\nu}\delta_{m+n,0}
\end{aligned} \tag{7.41}$$

The presence of the Minkowski metric in these equations mean that the theory will still be plagued by negative norm states that we will have to remove.

The Super-Virasoro Algebra

The Virasoro operators generate what is known as a *super-Virasoro algebra*. There are some differences for the R sector and the NS sector, so we consider each independently.

NS SECTOR ALGEBRA

For the NS sector, the following relations are satisfied:

$$[L_n, L_m] = (n - m)L_{n+m} + \frac{c}{12}(n^3 - n)\delta_{n+m,0} \tag{7.42}$$

$$[L_n, G_r] = \frac{1}{2}(n - 2r)G_{n+r} \tag{7.43}$$

$$\{G_r, G_s\} = 2L_{r+s} + \frac{c}{12}(4r^2 - 1)\delta_{r+s,0} \tag{7.44}$$

The central charge is related to the space-time dimension by $c = D + D/2$. Let $|\psi\rangle$ be a physical state in the NS sector. The NS sector super-Virasoro constraints are

$$(L_0 - a_{NS})|\psi\rangle = 0 \tag{7.45}$$

$$L_n|\psi\rangle = 0 \quad n > 0 \tag{7.46}$$

$$G_r|\psi\rangle = 0 \quad r > 0 \tag{7.47}$$

Here, following the quantization of the bosonic string, a_{NS} is a normal-ordering constant. The open string mass formula is taken by setting $L_0 = a_{NS}$, which gives

$$m^2 = \frac{1}{\alpha'}(N - a_{NS}) \tag{7.48}$$

Where the number operator is

$$N = \sum_{n=1}^{\infty} \alpha_{-n} \cdot \alpha_n + \sum_{r=1/2}^{\infty} r b_{-r} \cdot b_r \tag{7.49}$$

R SECTOR ALGEBRA

In the R sector, the commutation and anticommutation relations are

$$[L_m, L_n] = (m-n)L_{m+n} + \frac{D}{8}m^3\delta_{m+n,0} \tag{7.50}$$

$$[L_m, F_n] = \left(\frac{m}{2} - n\right)F_{m+n} \tag{7.51}$$

$$\{F_m, F_n\} = 2L_{m+n} + \frac{D}{2}m^2\delta_{m+n,0} \tag{7.52}$$

The conditions on the physical states are

$$(L_0 - a_R)|\psi\rangle = 0 \tag{7.53}$$

$$L_n|\psi\rangle = 0 \quad n > 0 \tag{7.54}$$

$$F_m|\psi\rangle = 0 \quad m \geq 0 \tag{7.55}$$

Here, a_R is the normal-ordering constant for the R sector.

EXAMPLE 7.6
Deduce that $a_R = 0$.

SOLUTION
We start with the anticommutation relation satisfied by the F_m:

$$\{F_m, F_n\} = 2L_{m+n} + \frac{D}{2}m^2\delta_{m+n,0}$$

Notice that if $m = n = 0$, we obtain

$$\{F_0, F_0\} = F_0 F_0 + F_0 F_0 = 2F_0^2 = 2L_0$$
$$\Rightarrow L_0 = F_0^2$$

The F_m annihilate physical states $|\psi\rangle$. Therefore,

$$F_0|\psi\rangle = 0$$

From this we obtain, by acting on the equation with F_0, the relation

$$F_0\left(F_0|\psi\rangle\right) = F_0^2|\psi\rangle = 0$$
$$\Rightarrow L_0|\psi\rangle = 0$$

But we know that $(L_0 - a_R)|\psi\rangle = 0$. Hence,

$$0 = (L_0 - a_R)|\psi\rangle = L_0|\psi\rangle - a_R|\psi\rangle = -a_R|\psi\rangle$$
$$\Rightarrow a_R = 0$$

The Open String Spectrum

Now let's examine the states of the string. We will look at states of the open string in this chapter. We must consider the NS and R sectors independently. Working in the NS sector first, the ground state is $|0,k\rangle_{NS}$ and it is annihilated by the modes

$$\alpha_n^i|0,k\rangle_{NS} = b_r^i|0,k\rangle_{NS} = 0 \tag{7.56}$$

where $n, r > 0$. The zero mode α_0^μ as discussed in the bosonic string case is a momentum operator:

$$\alpha_0^\mu|0,k\rangle_{NS} = \sqrt{2\alpha'}|0,k\rangle_{NS} \tag{7.57}$$

It can be shown that the normal-ordering constant in the NS sector is

$$a_{NS} = \frac{1}{2} \tag{7.58}$$

Using this we can find the mass of the ground state, which is

$$m^2 = -\frac{1}{2\alpha'} \tag{7.59}$$

Once again, we have a state with $m^2 < 0$, so the theory still contains a tachyon state. We will see later that we can get rid of the tachyon state in the superstring theory. The ground state in the NS sector is a unique spin-0 state. To find massive states, we progressively act on the state with negative mode oscillators.

Next we consider the R sector, which describes space-time fermions in the open string case. The ground state is annihilated by

$$\alpha_m^\mu |0,k\rangle_R = d_m^\mu |0,k\rangle_R = 0 \tag{7.60}$$

for $m > 0$. The zero mode d_0^μ is actually a Dirac operator. That is,

$$d_0^\mu = \Gamma^\mu \tag{7.61}$$

We will see below that the critical space-time dimension is 10, so the states in the R sector are 10-dimensional spinors. The ground state satisfies the massless Dirac wave equation. In our notation, this is written in the following way, recalling that the momentum operator is α_0^μ:

$$\alpha_0 \cdot d_0 |0,k\rangle_R = 0 \tag{7.62}$$

From Eq. (7.61), we deduce that the ground state in the R sector is a massless Dirac spinor in 10 dimensions.

GSO Projection

In the previous section, we saw that the theory still has a major problem—it admits an imaginary mass or tachyon state. This indicates that the vacuum is unstable. We can rid the theory of the tachyon state, however, giving superstring theory a major advantage over bosonic string theory (aside from bringing fermions into the picture). This is done using *GSO projection.*

GSO projection reduces the number of states in the theory, and rids it of unwanted problems like the tachyon state. In the NS sector, we keep states with an odd number of fermion excitations and reject states with an even number of fermion excitations. This is done by defining a *fermion number operator*:

$$F = \sum_{r>0} b_{-r} \cdot b_r \tag{7.63}$$

Then we define a *parity operator* given by

$$P_{NS} = \frac{1}{2}[1 - (-1)^F] \tag{7.64}$$

The parity operator determines the states that we can have in the theory. Notice that if $F = 0, \Rightarrow P_{NS} = 0.$ Only half integer values of the number operator $N = \sum_{n=1}^{\infty} \alpha_{-n} \cdot \alpha_n + \sum_{r=1/2} rb_{-r} \cdot b_r \rightarrow \sum_{r=1/2} rb_{-r} \cdot b_r$ are allowed, giving a mass spectrum for the NS sector:

$$m^2 = 0, \frac{1}{\alpha'}, \frac{2}{\alpha'}, \cdots \qquad (7.65)$$

This means that the spin-0 ground state of the NS sector is now *massless*. The tachyon state has been removed from the theory.

In the R sector, we define the *Klein operator* which is given by

$$(-1)^F = \pm \Gamma^{11} \qquad (7.66)$$

Here,

$$\Gamma^{11} = \Gamma^0 \Gamma^1 \cdots \Gamma^9 \qquad (7.67)$$

is a 10-dimensional chirality operator. It acts on spinors ψ according to

$$\Gamma^{11}\psi = \pm\psi \qquad (7.68)$$

That is, states have positive or negative chirality. Weyl spinors are states with definite chirality, and states can be projected into spinors with opposite space-time chirality using the operator

$$P_{\pm} = \frac{1}{2}(1 \pm \Gamma_{11}) \qquad (7.69)$$

Critical Dimension

We will not pursue light-cone quantization in this chapter, but if that procedure is used the number of space-time dimensions is easily extracted. One obtains a relation for the Lorentz generators M^i:

$$[M^{-i}, M^{-j}] = -\frac{1}{(p^+)^2} \sum_{n=1}^{\infty} (\alpha_{-n}^i \alpha_n^j - \alpha_{-n}^j \alpha_n^i)(\Delta_n - n) \qquad (7.70)$$

where

$$\Delta_n = n\left(\frac{D-2}{8}\right) + \frac{1}{n}\left(2a_{NS} - \frac{D-2}{8}\right) \tag{7.71}$$

In order to maintain Lorentz invariance, we must have $[M^{-i}, M^{-j}] = 0$. This can only be true if the first term on the right-hand side of Eq. (7.71) is n and the second term vanishes. This implies that

$$\frac{D-2}{8} = 1$$
$$\Rightarrow D = 10 \tag{7.72}$$

So, we see that

- Lorentz invariance requires us to take the critical space-time dimension to be 10 (9 space and 1 time dimension) in superstring theory.

Using Eq. (7.72), we can deduce the value of the normal-ordering constant:

$$2a_{NS} - \frac{D-2}{8} = 0 \qquad D = 10 \Rightarrow a_{NS} = \frac{1}{2}$$

Summary

In this chapter, we made the first attempt to introduce fermions to string theory. This was done by adding supersymmetry as a global symmetry on the worldsheet. The conserved current and supercurrent was derived. Next, we wrote down the super-Virasoro algebra and determined how physical states behave in the theory, and the spectrum of the open string was described including the two sectors, the NS and R sectors which give rise to bosonic and fermionic states, respectively. Using GSO projection, one can remove unwanted states like the Tachyon from the theory. Finally, we showed how Lorentz invariance forces us to take the critical dimension to be 10.

Quiz

1. Compute δS_F to arrive at the equations of motion [Eq. (7.8)].
2. Using $S = -T/2 \int d^2\sigma \, (\partial_\alpha X^\mu \partial^\alpha X_\mu - i\bar{\psi}^\mu \rho^\alpha \partial_\alpha \psi_\mu)$, consider an infinitesimal Lorentz transformation $X^\mu \to \omega_{\mu\nu} X^\nu$. Find the conserved current associated with the fermionic part of the lagrangian. (Hint: The Majorana spinors used here transform as vectors under Lorentz transformations.)

3. A continuation of Prob. 2. What is the total conserved current? (Hint: $\omega_{\mu\nu}$ is antisymmetric.)

4. For the worldsheet supercurrent $J^{\mu}_{\alpha} = 1/2 \rho^{\beta} \rho_{\alpha} \psi^{\mu} \partial_{\beta} X_{\mu}$, calculate $\rho^{\alpha} J_{\alpha}$.

5. Using Eq. (7.14), find δL_{B} given $S = -T/2 \int d^{2}\sigma \, (\partial_{\alpha} X^{\mu} \, \partial^{\alpha} X_{\mu} - i \bar{\psi}^{\mu} \rho^{\alpha} \partial_{\alpha} \psi_{\mu})$.

6. Calculate $\partial_{+} T_{--}$.

7. Find $\left\{ (-1)^{F}, b^{\mu}_{r} \right\}$.

8. Find $\{ \Gamma^{\mu}, \Gamma^{11} \}$.

9. Calculate $(\Gamma^{11})^{2}$.

10. Characterize the states $\alpha^{\mu}_{-1} |0\rangle_{NS}$ and $d^{\mu}_{-1} |0\rangle_{R}$.

CHAPTER 8

Compactification and T-Duality

In this chapter we introduce two important concepts. The first, *compactification*, involves compactifying one of the extra spatial dimensions into a circle of radius R. The second, *T-duality*, allows us to relate a theory with compactified extra dimension of radius R to one with compactified extra dimension of radius α'/R.

Compactification of the 25th Dimension

For simplicity, we consider the theory of the bosonic string which means we are back to working with 26 space-time dimensions for the moment. The dimension X^0 is timelike while X^1, ..., X are spatial dimensions. We imagine that one of those spatial dimensions, typically chosen to be X^{25} is curled up into a circle of radius R. We wish to study how this compactification affects closed strings. As we will see, it has some interesting consequences.

Previously a closed string was constrained by the following periodic boundary condition:

$$X^\mu(\sigma,\tau) = X^\mu(\sigma+2\pi,\tau) \tag{8.1}$$

This boundary condition was stated with the implicit assumption where the string was moving in a space-time with noncompact dimensions. Now let's modify the situation. As stated above, we are going to let the 25th dimension be a circle with radius R. This changes the boundary condition in Eq. (8.1) as follows, but only for X^{25}:

$$X^{25}(\sigma+2\pi,\tau) = X^{25}(\sigma,\tau) + 2\pi nR \tag{8.2}$$

The interesting thing about Eq. (8.2) is that now the string will have *winding states.* Simply put, the string can wind around the compactified dimension any number of times. For all other dimensions $\mu \neq 25$, Eq. (8.1) still holds.

The number n in Eq. (8.2) is called the *winding number.* Using the winding number we can define the *winding w* as

$$w = \frac{nR}{\alpha'} \tag{8.3}$$

We are going to see in a moment that the winding is actually a type of momentum. The periodic boundary condition in Eq. (8.2) can be written in terms of the winding as

$$X^{25}(\sigma+2\pi,\tau) = X^{25}(\sigma,\tau) + 2\pi\alpha'w \tag{8.4}$$

Now let's take a closer look at the boundary condition by seeing how it affects left-moving and right-moving modes. This will demonstrate the fact that the winding is a kind of momentum. First recall that $X^\mu(\sigma,\tau) = X_L^\mu(\sigma,\tau) + X_R^\mu(\sigma,\tau)$. The left- and right-moving modes can be written as follows:

$$X_L^{25}(\sigma,\tau) = \frac{1}{2}x_L^{25} + \frac{\alpha'}{2}p_L^{25}(\tau+\sigma) + i\sqrt{\frac{\alpha'}{2}}\sum_{n\neq 0}\frac{\bar\alpha_n}{n}e^{-in(\tau+\sigma)} \tag{8.5}$$

$$X_R^{25}(\sigma,\tau) = \frac{1}{2}x_R^{25} + \frac{\alpha'}{2}p_R^{25}(\tau-\sigma) + i\sqrt{\frac{\alpha'}{2}}\sum_{n\neq 0}\frac{\alpha_n}{n}e^{-in(\tau-\sigma)} \tag{8.6}$$

where we have made the identifications $\alpha_0^{25} = (\alpha'/2)^{1/2} p_L^{25}$ and $\bar\alpha_0^{25} = (\alpha'/2)^{1/2} p_R^{25}$. Adding Eqs. (8.5) and (8.6) together (while ignoring the oscillator contributions) we get

$$X^{25}(\sigma,\tau) = x^{25} + \frac{\alpha'}{2}\left(p_L^{25} + p_R^{25}\right)\tau + \frac{\alpha'}{2}\left(p_L^{25} - p_R^{25}\right)\sigma + \text{ modes} \tag{8.7}$$

The total center of mass momentum of the string is

$$p^{25} = p_L^{25} + p_R^{25} \tag{8.8}$$

Along the compactified dimension, the string acts like a particle moving on a circle. The momentum is quantized according to

$$p^{25} = \frac{K}{R} \tag{8.9}$$

where K is an integer called the *Kaluza-Klein excitation number*. This is an important result—without the compactified dimension, the center of mass momentum of the string is continuous. Compactifying a dimension quantizes the center of mass momentum along that dimension.

Looking at Eq. (8.7) then, the first term involving the momenta is the total center of mass momentum of the string. We call this the *momentum mode*. The second term, however, also involves momentum. In fact this term is the *winding mode* of the string, which satisfies

$$\frac{\alpha'}{2}\left(p_L^{25} - p_R^{25}\right) = nR \tag{8.10}$$

Looking at Eq. (8.3), we see that the winding w can be defined in terms of the momentum of the left- and right-moving modes as

$$w = \frac{nR}{\alpha'} = \frac{1}{2\alpha'}\left(p_L^{25} - p_R^{25}\right) \tag{8.11}$$

Modified Mass Spectrum

Compactifying a dimension will lead to a modified mass spectrum. To obtain the mass spectrum for the state with a compactified dimension, let us begin with the Virasoro operators. Recall that

$$L_0 = \frac{\alpha'}{4} p_R^\mu p_{R\mu} + \sum_{n=1}^{\infty} \alpha_{-n} \cdot \alpha_n \tag{8.12}$$

Note the repeated index which is an upper and lower index on the first term in the right—so we have an implied sum. Here the index μ ranges over the entire space-time, that is, $\mu = 0, ..., 2$. Now let's write L_0 in such a way that we peel of the $\mu = 25$ term. Then

$$L_0 = \frac{\alpha'}{4} p_R^{25} p_R^{25} + \frac{\alpha'}{4} \sum_{\mu=0}^{24} p_R^{\mu} p_{R\mu} + \sum_{n=1}^{\infty} \alpha_{-n} \cdot \alpha, \tag{8.13}$$

Similarly we can write

$$\bar{L}_0 = \frac{\alpha'}{4} p_L^{25} p_L^{25} + \frac{\alpha'}{4} \sum_{\mu=0}^{24} p_L^{\mu} p_{L\mu} + \sum_{n=1}^{\infty} \bar{\alpha}_{-n} \cdot \bar{\alpha}, \tag{8.14}$$

With a single compactified dimension, the Kaluza-Klein excitations on X^{25} are considered to be *distinct particles*. Hence we can write down the mass operator as a mass term in the 25 noncompactified dimensions. That is,

$$m^2 = -\sum_{\mu=0}^{24} p^{\mu} p_{\mu} \tag{8.15}$$

Now, you should recognize the sums $\sum_{n=1}^{\infty} \alpha_{-n} \cdot \alpha_n$ and $\sum_{n=1}^{\infty} \bar{\alpha}_{-n} \cdot \bar{\alpha}_n$ as the number operators N_R and N_L. Using this fact together with Eq. (8.15) allows us to write the Virasoro operators as

$$L_0 = \frac{\alpha'}{4} p_R^{25} p_R^{25} - \frac{\alpha'}{4} m^2 + N_R \tag{8.16}$$

$$\bar{L}_0 = \frac{\alpha'}{4} p_L^{25} p_L^{25} - \frac{\alpha'}{4} m^2 + N_L \tag{8.17}$$

Now we can utilize the mass-shell constraint. This is the condition that $L_0 - 1$ and $\bar{L}_0 - 1$ annihilate physical states $|\psi\rangle$:

$$(L_0 - 1)|\psi\rangle = 0 \tag{8.18}$$

$$(\bar{L}_0 - 1)|\psi\rangle = 0 \tag{8.19}$$

The conditions in Eqs. (8.18) and (8.19) imply that $L_0 = 1$ and $\bar{L}_0 = 1$. Applying the first condition to Eq. (8.16) we get

$$L_0 = 1$$

$$\Rightarrow 1 = \frac{\alpha'}{4} p_R^{25} p_R^{25} - \frac{\alpha'}{4} m^2 + N_R$$

$$\Rightarrow \frac{\alpha'}{2} m^2 = \frac{\alpha'}{2} p_R^{25} p_R^{25} + 2N_R - 2 \tag{8.20}$$

Similarly, using $\bar{L}_0 = 1$ together with Eq. (8.17) we obtain

$$\frac{\alpha'}{2}m^2 = \frac{\alpha'}{2}p_L^{25}p_L^{25} + 2N_L - 2 \qquad (8.21)$$

Now using Eq. (8.8) together with Eqs. (8.9) and (8.10) we can write

$$p_L^{25} = \frac{nR}{\alpha'} + \frac{K}{R} \qquad (8.22)$$

Which of course allows us to compute

$$\left(p_L^{25}\right)^2 = \left(\frac{nR}{\alpha'} + \frac{K}{R}\right)^2 = \left(\frac{nR}{\alpha'}\right)^2 + \left(\frac{K}{R}\right)^2 + 2\frac{nK}{\alpha'} \qquad (8.23)$$

and similarly $p_R^{25} = (K/R) - (nR/\alpha')$ so that

$$\left(p_R^{25}\right)^2 = \left(\frac{K}{R} - \frac{nR}{\alpha'}\right)^2 = \left(\frac{nR}{\alpha'}\right)^2 + \left(\frac{K}{R}\right)^2 - 2\frac{nK}{\alpha'} \qquad (8.24)$$

This allows us to obtain the sum and difference formulas:

$$\left(p_L^{25}\right)^2 + \left(p_R^{25}\right)^2 = 2\left[\left(\frac{nR}{\alpha'}\right)^2 + \left(\frac{K}{R}\right)^2\right] \qquad (8.25)$$

$$\left(p_L^{25}\right)^2 - \left(p_R^{25}\right)^2 = 4\frac{nK}{\alpha'} \qquad (8.26)$$

Using Eqs. (8.25) and (8.26) we can add Eqs. (8.20) and (8.21) to obtain

$$\alpha'm^2 = \left(\frac{nR}{\alpha'}\right)^2 + \left(\frac{K}{R}\right)^2 + 2(N_R + N_L) - 4 \qquad (8.27)$$

and subtracting Eqs. (8.21) from (8.20) gives

$$N_R - N_L = nK \qquad (8.28)$$

So notice we have extra terms in the formulas for mass [Eq. (8.27)] and the *level matching condition* [Eq. (8.28)] as compared to the formulas introduced for the bosonic string in Chap. 2. The extra terms are due to two components, the Kaluza-Klein excitations and the winding states of the string. The Kaluza-Klein excitations can be regarded as particles and so cannot be thought of as due to strings in a general sense. However the winding excitations can only come from strings, because only strings can wrap around a compactified extra dimension.

Now let's look at the mass formula in Eq. (8.27) together with the relations for the momenta in Eqs. (8.22) and (8.25). Our task here is to consider limiting behavior. First we consider the case where $R \rightarrow \infty$. In this limit, the momentum goes to the continuum limit and the Kaluza-Klein excitations disappear. It is simple to show that $p_L = p_R$ and hence the winding state

$$w = \frac{1}{2}\left(p_L^{25} - p_R^{25}\right) \rightarrow 0 \qquad \text{as } R \rightarrow \infty$$

The center of mass momentum Eq. (8.8) returns to the nonquantized, continuous momentum of the noncompactified case.

Now let's think about the opposite limit where $R \rightarrow 0$. You might also expect that this is like returning to the noncompactified case. After all, taking the limit $R \rightarrow 0$ is like making the extra dimension go away. In quantum field theory we might expect the fields to completely decouple from that unseen extra dimension. However, things don't quite work this way in string theory.

As $R \rightarrow 0$, we find that the Kaluza-Klein modes become infinitely massive and decouple from the theory. Since these can be regarded as particle states, maybe this isn't so surprising. What's left behind for the center of mass momentum are the winding states. First note that as $R \rightarrow 0$ we obtain

$$p_R = -p_L$$

Hence $p^{25} \rightarrow 0$ but the winding term behaves in the following way:

$$w = \frac{1}{2}\left(p_L^{25} - p_R^{25}\right) \rightarrow p_L^{25} \qquad \text{as } R \rightarrow 0$$

(or to $-p_R^{25}$ if you like). Now the winding states, rather than the momentum states, form a continuum of states. This should not be so surprising, as $R \rightarrow 0$ the circle gets smaller and smaller. So it gets easier and easier to wrap a string around it—that is, it costs less energy. When the circle is very small it doesn't require a lot of energy to wrap the string around it.

So you see as the radius gets very large or very small there is a trade-off between winding states and momentum. This trade-off leads us to a discussion of T-duality, the topic of the next section.

T-Duality for Closed Strings

T-duality is a symmetry which exists between different string theories. This symmetry relates small distances in one theory to large distances in another, seemingly different theory and shows that the two theories are in fact the same theory expressed from different viewpoints. This is an important recognition; before T-duality was discovered it was believed that there were five different string theories, when in fact they were all different versions of the same theory that could be related to one another by transformations or dualities. One can transform between small and large distances when considering the compactified dimension in one theory, and arrive at another dual theory. This is the essence of T-duality. We will see later that other dualities exist in string theory as well.

T-duality relates type IIA and type IIB string theories, as well as the heterotic string theories. It applies to the type of compactification that we have been studying in this chapter, namely the compactification of a spatial dimension to a circle of radius R. The transformation that is used in T-duality is to transform the radius to a new large radius R' which is defined by the exchange

$$R' \leftrightarrow \frac{\alpha'}{R} \tag{8.29}$$

The T-duality transformation also exchanges winding states characterized by a winding number n with high-momentum states in the other theory (Kaluza-Klein excitations). That is,

$$n \leftrightarrow K \tag{8.30}$$

The symmetry of T-duality, described by these exchanges, makes its appearance in the mass formula [Eq. (8.27)], which we reproduce here:

$$\alpha' m^2 = \left(\frac{nR}{\alpha'}\right)^2 + \left(\frac{K}{R}\right)^2 + 2(N_R + N_L) - 4$$

Now exchange $R' \leftrightarrow \alpha'/R$ and $n \leftrightarrow K$, then:

$$\frac{K}{R} \rightarrow \frac{n}{\left(\alpha'/R'\right)} = \frac{nR'}{\alpha'} \tag{8.31}$$

We also have:

$$\frac{nR}{\alpha'} \rightarrow \frac{K\left(\alpha'/R'\right)}{\alpha'} = \frac{K}{R'} \tag{8.32}$$

So we see that the mass formula Eq. (8.27) is *invariant* under the exchange $R' \leftrightarrow \alpha'/R$ and $n \leftrightarrow K$. It assumes the form

$$\alpha' m^2 = \left(\frac{nR'}{\alpha'}\right)^2 + \left(\frac{K}{R'}\right)^2 + 2(N_R + N_L) - 4$$

That is, it keeps the same form but now with the new radius R'. That's the math of the transformation. The physics is that if we started with a theory with a small compactified dimension R, we have transformed to a *dual* theory with a large extra dimension R'. What this means for the string is that a string in a type IIA theory (with small compactified dimension), which winds around the small compact dimension (with winding states) is dual to a string in type IIB theory (with the dimension transformed to a large dimension of radius R'), which has *momentum* along that dimension. Each time the string in type IIA theory winds around the compact dimension, this corresponds to increasing the momentum in type IIB theory by one unit.

Now let's examine how p_L^{25} and p_R^{25}, and by extension α_0 and $\bar{\alpha}_0$, transform under this symmetry. Recall Eq. (8.22) that states

$$p_L^{25} = \frac{nR}{\alpha'} + \frac{K}{R}$$

Now exchange $R' \leftrightarrow \alpha'/R$ and $n \leftrightarrow K$. We find

$$p_L^{25} \rightarrow \frac{K}{\alpha'}\left(\frac{\alpha'}{R'}\right) + \frac{n}{\left(\alpha'/R'\right)} = \frac{K}{R'} + \frac{nR'}{\alpha'}$$

So, p_L^{25} maintains the same form under the transformation—that is, it is invariant. We can indicate this by writing $p_L^{25} = p_L^{25}$. Now consider how p_R^{25} transforms under the exchange $R' \leftrightarrow \alpha'/R$ and $n \leftrightarrow K$:

$$p_R^{25} = \frac{K}{R} - \frac{nR}{\alpha'}$$

$$\Rightarrow p_R^{25} \rightarrow \frac{n}{\left(\alpha'/R'\right)} - \frac{K}{\alpha'}\left(\frac{\alpha'}{R'}\right) = \frac{nR'}{\alpha'} - \frac{K}{R'}$$

That is, $p_R^{25} = -p_R^{25}$. This tells us that the exchange $R' \leftrightarrow \alpha'/R$ and $n \leftrightarrow K$ is equivalent to transformations applied to the zero-modes:

$$\alpha_0^{25} \rightarrow \alpha_0^{25} \quad \text{and} \quad \bar{\alpha}_0^{25} \rightarrow -\bar{\alpha}_0^{25} \qquad (8.33)$$

We can summarize this as follows:

- A state (K, n) in a theory with radius R is transformed into a state (n, K) in a theory with radius $R' = \alpha'/R$.
- The two states have the same mass. That is, $m^2(K, n, R) = m^2(n, K, R')$.
- The number operators are unchanged under a T-duality transformation.

The transformation not only preserves mass, the entire theory with compactified dimension of radius R is mapped to a theory of radius $R' = \alpha'/R$ using a T-duality transformation. Let ~ denote quantities in the dual theory. A T-duality transformation maps the compactified coordinates to those in the dual theory as follows:

$$X_L^{25}(\tau + \sigma) \rightarrow \tilde{X}_L^{25}(\tau + \sigma)$$
$$X_R^{25}(\tau + \sigma) \rightarrow -\tilde{X}_R^{25}(\tau + \sigma)$$

$$(8.34)$$

Then using $X^{25} = X_L^{25} + X_R^{25}$ the relation of the coordinates in the dual theory $\tilde{X}^{25} = \tilde{X}_L^{25} + \tilde{X}_R^{25}$ can be written as

$$\tilde{X}^{25} = X_L^{25} - X_R^{25}$$

Furthermore we have the relations

$$\partial_+ \tilde{X}_{25} = \partial_+ X_{25}$$
$$\partial_- \tilde{X}_{25} = -\partial_- X_{25}$$

However the physical content of the theory is unchanged because $\partial_+ X_R^{25} = \partial_- X_L^{25} = 0$. The winding number and Kaluza-Klein excitation in the theory with compactified dimension and its dual are related according to

$$K = \tilde{n} \quad \text{and} \quad n = \tilde{K}$$

A T-duality transformation also maps all the modes as

$$\alpha_n^{25} = -\tilde{\alpha}_n^{25}$$
$$\bar{\alpha}_n^{25} = \tilde{\bar{\alpha}}_n^{25}$$

Open Strings and T-Duality

The situation is a little different when considering compactification in the open string case. This is because open strings cannot wind around the compact dimension. We summarize this by saying:

- Open strings do not have winding modes when a dimension is compactified into a circle of radius R. Hence they have winding number $n = 0$.

Let's quickly review a couple of facts about open strings in bosonic string theory. In order to satisfy Poincaré invariance, we choose Neumann boundary conditions for the open string:

$$\frac{\partial X^\mu}{\partial \sigma} = 0 \qquad \text{for } \sigma = 0, \pi \tag{8.35}$$

The modal expansion for the open string with Neumann boundary conditions is given by

$$X^\mu(\sigma, \tau) = x_0^\mu + 2\alpha' p_0^\mu + i\sqrt{2\alpha'} \sum_{n \neq 0} \frac{\alpha_n^\mu}{n} e^{-in\tau} \cos n\sigma \tag{8.36}$$

We can write left-moving and right-moving modes for the open string as

$$X_L^\mu(\tau + \sigma) = \frac{x_0^\mu + \tilde{x}_0^\mu}{2} + \alpha' p_0^\mu(\tau + \sigma) + i\sqrt{\frac{\alpha'}{2}} \sum_{n \neq 0} \frac{\alpha_n^\mu}{n} e^{-in(\tau + \sigma)}$$

$$X_R^\mu(\tau - \sigma) = \frac{x_0^\mu - \tilde{x}_0^\mu}{2} + \alpha' p_0^\mu(\tau - \sigma) + i\sqrt{\frac{\alpha'}{2}} \sum_{n \neq 0} \frac{\alpha_n^\mu}{n} e^{-in(\tau - \sigma)} \tag{8.37}$$

Note that x_0^{25} will be the position coordinate along the compactified dimension. Here we have added and subtracted $\tilde{x}_0^\mu$, which is the coordinate of the compactified dimension in the dual space.

We can go through the compactification procedure by simply applying the T-duality transformation (which is why we have written the open string modes in terms of left movers and right movers). We let

$$X_L^{25} \to X_L^{25} \qquad \text{and} \qquad X_R^{25} \to -X_R^{25}$$

This means that for X^{25} we have:

$$\tilde{X}_L^{25}(\tau+\sigma) = X_L^{25}(\tau+\sigma) = \frac{x_0^{25} + \tilde{x}_0^{25}}{2} + \alpha' p_0^{25}(\tau+\sigma) + i\sqrt{\frac{\alpha'}{2}}\sum_{n\neq0}\frac{\alpha_n^{25}}{n}e^{-in(\tau+\sigma)}$$

$$\tilde{X}_R^\mu(\tau-\sigma) = -X_R^\mu(\tau-\sigma) = \frac{-x_0^{25} + \tilde{x}_0^{25}}{2} + \alpha' p_0^{25}(\tau-\sigma) - i\sqrt{\frac{\alpha'}{2}}\sum_{n\neq0}\frac{\alpha_n^{25}}{n}e^{-in(\tau-\sigma)}$$

$$(8.38)$$

Adding together to get the total mode expansion gives

$$\tilde{X}^{25} = \tilde{x}_0^{25} + \alpha' p_0^{25}\sigma + i\sqrt{\frac{\alpha'}{2}}\sum_{n\neq0}\frac{\alpha_n^{25}}{n}\left[e^{-in(\tau+\sigma)} - e^{-in(\tau-\sigma)}\right]$$

Now using Euler's famous formula:

$$e^{-in(\tau+\sigma)} - e^{-in(\tau-\sigma)} = 2i\left[\frac{e^{-in(\tau+\sigma)} - e^{-in(\tau-\sigma)}}{2i}\right]$$

$$= -2ie^{-in\tau}\left[\frac{e^{in\sigma} - e^{-in\sigma}}{2i}\right] = -2ie^{-in\tau}\sin n\sigma$$

This means that the mode expansion can be written as

$$\tilde{X}^{25} = \tilde{x}_0^{25} + \alpha' p_0^{25}\sigma + \sqrt{2\alpha'}\sum_{n\neq0}\frac{\alpha_n^{25}}{n}e^{-in\tau}\sin n\sigma$$

$$= \tilde{x}_0^{25} + \alpha'\frac{K}{R}\sigma + \sqrt{2\alpha'}\sum_{n\neq0}\frac{\alpha_n^{25}}{n}e^{-in\tau}\sin n\sigma$$

Now we can analyze this expression to discover the properties of open strings in the dual theory. The first item to notice is:

- The expression for $\tilde{X}^{25}$ has no linear terms that contain the worldsheet time coordinate τ. Physically, This means that the dual string has no momentum in the 25th dimension.

- If the string carries no momentum for $\mu = 25$, it must be fixed. What does a fixed vibrating string do? The motion is oscillatory.

- Notice that the expansion contains a $\sin n\sigma$ term, which of course satisfies $\sin n\sigma = 0$ at $\sigma = 0, \pi$.

The last point is particularly important. Recall that Dirichlet boundary conditions on the string are

$$X^\mu\big|_{\sigma=0} = X^\mu\big|_{\sigma=\pi} = 0$$

Looking at the expression for the dual field, notice that

$$\tilde{X}^{25}(\sigma = 0, \tau) = \tilde{x}_0^{25}$$

At $\sigma = \pi$ we have

$$\tilde{X}^{25}(\sigma = \pi, \tau) = \tilde{x}_0^{25} + \alpha'\pi\frac{K}{R} = \tilde{x}_0^{25} + 2K\pi R'$$

Hence,

$$\tilde{X}^{25}(\sigma = \pi, \tau) - \tilde{X}^{25}(\sigma = 0, \tau) = 2K\pi R'$$

This tells us that the dual string winds around the dual dimension of radius R' with winding number K.

Summarizing

- T-duality transforms Neumann boundary conditions into Dirichlet boundary conditions.
- T-duality transforms Dirichlet boundary conditions into Neumann boundary conditions.
- T-duality transforms a bosonic string with momentum but no winding into a string with winding but no momentum.
- For the dual string, the string endpoints are restricted to lie on a 25-dimensional hyperplane in space-time.
- The endpoints of the dual string can wind the circular dimension an integer number of times given by K.

D-Branes

The hyperplane that the open string is attached to carries special significance. A D-brane is a hypersurface in space-time. In the examples worked out in this chapter, it is a hyperplane with 24 spatial dimensions. The dimension which has been excluded in this example is the dimension which has been compactified. The D is short for *Dirichlet* which refers to the fact that the open strings in the theory have endpoints that satisfy Dirichlet boundary conditions. In English this means that the *endpoints of an open string are attached to a D-brane.*

A D-brane can be classified by the number of spatial dimensions it contains. A point is a zero-dimensional object and therefore is a D0-brane. A line, which is a one-dimensional object is a D1-brane (so strings can be thought of as D1-branes). Later we will see that the physical world of three spatial dimensions and one time dimension that we can perceive directly is a D3-brane contained in the larger world of 11-dimensional hyperspace. In the example studied in this chapter, we considered a D24-brane, with one spatial dimension compactified that leaves 24 dimensions for the hyperplane surface.

Using the procedure outlined here, other dimensions can be compactified. If we choose to compactify n dimensions then that leaves behind a D(25-n)-brane. The procedure outlined here is essentially the same in superstring theory, but in that case compactifying n dimensions gives us a D(9-n)-brane. Note that:

- The ends of an open string are free to move in the noncompactified directions—including time. So in bosonic theory, if we have compactified n directions, the endpoints of the string are free to move in the other $1 + (25-n)$ directions. In superstring theory, the endpoints will be free to move in the other $1 + (9-n)$ directions. In the example considered in this chapter where we compactified 1 dimension in bosonic string theory, the end points of the string are free to move in the other $1 + 24$ dimensions.

We can consider the existence of D-branes to be a consequence of the symmetry of T-duality. The number, types, and arrangements of D-branes restrict the open string states that can exist. We will have more to say about D-branes and discuss T-duality in the context of superstrings in future chapters.

Summary

In this chapter we described *compactification* which involves taking a spatial dimension and compactifying it to a small circle of radius R. Going through this procedure, it was discovered that a symmetry emerges called *T-duality*, which relates theories with small R to equivalent theories with large R. An important consequence of T-duality was discovered when it was learned that open strings with Neumann boundary conditions are transformed into open strings with Dirichlet boundary conditions in the dual theory. The result is the endpoints of the string are fixed to a hyperplane called a D-brane.

Quiz

1. Translational invariance along σ leads to the condition $(L_0 - \bar{L}_0)|\psi\rangle = 0$ for physical states $|\psi\rangle$. Use this to find a relation between p_L^{25}, N_L, and p_R^{25}, N_R.

CHAPTER 9

Superstring Theory Continued

In Chap. 7, we took our first look at superstring theory by considering *supersymmetry on the worldsheet*. The result is the RNS superstring. We can learn a lot from this method but space-time supersymmetry is not manifest with this theory. In this chapter, we have a somewhat random collection of material on supersymmetry and superstrings that will give you a general overview of what these topics are about so that you can pursue more advanced treatments if desired. In short, we are going to do two things. First, we will deepen and extend our discussion of supersymmetry and superstrings, and then we will introduce a space-time supersymmetry approach. These are more advanced topics so we aren't going to go into great detail, and leave out a lot of important information. But our purpose here is to provide the reader with an introductory overview that exposes you to some of the basic ideas of superstring theory. A detailed study can be undertaken by reading any of the references listed at the end of the book. We hope this first exposure will make going through material in other textbooks a bit easier. The material in this chapter will not be necessary to understand D-branes or black hole physics as discussed in this book, so if you'd rather avoid it for now you can do so without much harm.

Superspace and Superfields

Adding space-time supersymmetry is going to involve a couple of things. Specifically:

- We will extend the coordinates to add a "supersymmetric partner" to the space-time coordinate x^μ. The result will be *superspace* defined by coordinates x^μ and θ_A.

- We will introduce a *superfield* which is a function of the superspace coordinates. The superfield will be added to the action to generate a supersymmetric theory.

With these points in mind let's first move ahead by describing the concept known as superspace. As noted above, the idea here is to add to the usual space-time coordinate x^0, x^1, ..., x^d by adding *fermionic* or *Grassman coordinates* θ_A. The index A used on the superspace or Grassman coordinates corresponds to the spinor index used on the spinors ψ_A^μ. Taking the case of worldsheet supersymmetry that we have discussed already, we had two component spinors, and so $A = 1, 2$.

Fermionic coordinates θ_A are also called Grassman coordinates because they satisfy an anticommution relation. That is,

$$\theta_A \theta_B + \theta_B \theta_A = 0 \tag{9.1}$$

Notice that this relation implies that $\theta_A \theta_A = \theta_A^2 = 0$. In the case of the worldsheet, the θ_A are super-worldsheet coordinates that are two component spinors:

$$\theta_A = \begin{pmatrix} \theta_- \\ \theta_+ \end{pmatrix}$$

To characterize superspace, we also need to understand how the fermionic coordinates behave with respect to normal space-time coordinates—this is encapsulated in commutation and anticommutation relations. Sticking to the worldsheet as an example, we denote the coordinates of the worldsheet by $\sigma^a = (\tau, \sigma)$. Since these are ordinary coordinates, they commute with themselves:

$$\sigma^a \sigma^b - \sigma^b \sigma^a = 0 \tag{9.2}$$

They also commute with the fermionic coordinates:

$$\sigma^a \theta_A - \theta_A \sigma^a = 0 \qquad (9.3)$$

So, what we've seen here is that supersymmetry doesn't just pair up bosons and fermions, it also enlarges the notion of space-time to pair up ordinary coordinates with fermionic coordinates, with superspace being characterized by the relations given by Eqs. (9.1) through (9.3). Now let's consider the notion of a *superfield*.

It is possible to define functions on superspace, meaning that we can introduce fields Y that are functions of space-time coordinates and fermionic coordinates. We can indicate this by writing

$$Y \equiv Y(\sigma, \theta)$$

for a given field Y. We call a field that is a function of superspace a *superfield*. A superfield can be introduced into the action to construct a supersymmetric theory.

Next, we introduce the *supercharge* which can also be called the *supersymmetry generator*. In the case of worldsheet supersymmetry, this is given by

$$Q_A = \frac{\partial}{\partial \bar{\theta}_A} - i(\rho^\alpha \theta)_A \partial_\alpha$$

We call the supercharge the supersymmetry generator because it generates supersymmetry transformations on superspace. That is, it acts on the coordinates as follows. Using the worldsheet example, the role of the space-time coordinates are played by $\sigma^\alpha = (\tau, \sigma)$. The supersymmetry generator acts on them as follows:

$$\bar{\varepsilon}Q\sigma^\alpha = \bar{\varepsilon}\left(\frac{\partial}{\partial\bar{\theta}} - i\rho^\beta\theta\partial_\beta\right)\sigma^\alpha$$

$$= \bar{\varepsilon}\left(\frac{\partial\sigma^\alpha}{\partial\bar{\theta}} - i\rho^\beta\theta\partial_\beta\sigma^\alpha\right) \qquad \text{(the } \sigma^\alpha \text{ coordinate is not a function of the fermion coordinate } \theta)$$

$$= -i\bar{\varepsilon}\rho^\beta\theta\delta^\alpha_\beta = -i\bar{\varepsilon}\rho^\alpha\theta$$

Hence we conclude that under a supersymmetry transformation

$$\sigma^\alpha \to \sigma^\alpha - i\bar{\varepsilon}\rho^\alpha\theta$$

This can also be written as $\sigma^\alpha \to \sigma^\alpha + i\bar{\theta}\rho^\alpha \varepsilon$. This is possible because of the properties of Grassman numbers together with $\{\rho^\alpha, \rho^\beta\} = 2\eta^{\alpha\beta}$ (remember that we have Majorana spinors that consist of anticommuting Grassman numbers). For readers new to the subject it's a good idea to write out the details, so let's take an aside to do so. We can reorder the term $\bar{\varepsilon}\rho^\alpha \theta$ in two ways. Do you remember how to reorder terms using the hermitian conjugate † in ordinary quantum mechanics? If so, then you will understand how to work with $\bar{\varepsilon}\rho^\alpha \theta$, but in this case we only use the transpose. We have

$$
\begin{aligned}
\bar{\varepsilon}\rho^\alpha \theta &= \varepsilon\rho^0 \rho^\alpha \theta = [(\rho^\alpha \theta)(\varepsilon\rho^0)]^T \\
&= \theta\rho^\alpha \rho^0 \varepsilon \\
&= \theta(2\eta^{\alpha 0} - \rho^0 \rho^\alpha)\varepsilon \\
&= 2\eta^{\alpha 0}\theta\varepsilon - \theta\rho^0 \rho^\alpha \varepsilon \\
&= 2\eta^{\alpha 0}\theta\varepsilon - \bar{\theta}\rho^\alpha \varepsilon
\end{aligned}
$$

But we can also write

$$
\begin{aligned}
\bar{\varepsilon}\rho^\alpha \theta &= \varepsilon\rho^0 \rho^\alpha \theta \\
&= \varepsilon(2\eta^{0\alpha} - \rho^\alpha \rho^0)\theta \\
&= 2\eta^{0\alpha}\varepsilon\theta - \varepsilon\rho^\alpha \rho^0 \theta \\
&= 2\eta^{0\alpha}\varepsilon\theta - [(\rho^0 \theta)(\varepsilon\rho^\alpha)]^T \\
&= 2\eta^{0\alpha}\varepsilon\theta - \theta\rho^0 \rho^\alpha \varepsilon = 2\eta^{0\alpha}\varepsilon\theta - \bar{\theta}\rho^\alpha \varepsilon
\end{aligned}
$$

Now, we use the fact that the metric is symmetric, so $\eta^{0\alpha} = \eta^{\alpha 0}$. Adding our two expressions, we get

$$
2\bar{\varepsilon}\rho^\alpha \theta = 2\eta^{0\alpha}(\theta\varepsilon + \varepsilon\theta) - 2\bar{\theta}\rho^\alpha \varepsilon
$$

We can write $\theta\varepsilon + \varepsilon\theta$ in terms of components and use the fact that the components are anticommuting Grassman numbers to show that the first term vanishes. First we have

$$
\theta\varepsilon = (\theta_- \quad \theta_+)\begin{pmatrix} \varepsilon_- \\ \varepsilon_+ \end{pmatrix} = \theta_- \varepsilon_- + \theta_+ \varepsilon_+
$$

We also have

$$
\varepsilon\theta = (\varepsilon_- \quad \varepsilon_+)\begin{pmatrix} \theta_- \\ \theta_+ \end{pmatrix} = \varepsilon_- \theta_- + \varepsilon_+ \theta_+
$$

Now, use the fact that the components are anticommuting to write

$$\varepsilon\theta = \varepsilon_-\theta_- + \varepsilon_+\theta_+ = -\theta_-\varepsilon_- - \theta_+\varepsilon_+ = -\theta\varepsilon$$

This means that $2\eta^{0\alpha}(\theta\varepsilon + \varepsilon\theta) = 0$, and so we are able to write $\bar\varepsilon\rho^\alpha\theta = -\bar\theta\rho^\alpha\varepsilon$.

Returning to the main thrust of our discussion, in general, space-time coordinates will transform as

$$x^\mu \to x^\mu - i\bar\varepsilon\gamma^\mu\theta$$

under a supersymmetry transformation, where γ^μ are the usual Dirac matrices. Now let's consider the action of the supersymmetry generator on the fermionic or super-worldsheet coordinates. We simply state the result which you can work out in the Chapter Quiz:

$$\delta\theta_A = \varepsilon_A$$

Hence, the supercoordinates transform as

$$\theta_A \to \theta_A + \varepsilon_A$$

A tool we will use to write down the action is the *supercovariant derivative*. This is given by

$$D_A = \frac{\partial}{\partial\bar\theta^A} + i(\rho^\alpha\theta)_A\partial_\alpha$$

A key property of the supercovariant derivative is that under a supersymmetry transformation, the supercovariant derivative of a superfield F, DF transforms the same way as F does.

Superfield for Worldsheet Supersymmetry

We will use the case of worldsheet supersymmetry to illustrate how to write down an action where supersymmetry is manifest. To do this, we start with the superfield:

$$Y^\mu(\sigma^\alpha,\theta) = X^\mu(\sigma^\alpha) + \bar\theta\psi^\mu(\sigma^\alpha) + \frac{1}{2}\bar\theta\theta B^\mu(\sigma^\alpha)$$

This expression is a general expression, it's a Taylor expansion of the superfield. Due to the anticommuting properties of Grassman variables $\bar{\theta}\theta$ is the highest-order term in the expansion. It can be shown that the equation of motion for B^μ is given by $B^\mu = 0$, so this is an auxiliary field that plays no role in the physics. The superfield transforms as

$$\delta Y^\mu = [\bar{\varepsilon}Q, Y^\mu] = \bar{\varepsilon}QY^\mu$$

Now recall Eq. (7.9), which gave the SUSY transformations of the boson and fermion fields X^μ and ψ^μ:

$$\delta X^\mu = \bar{\varepsilon}\psi^\mu$$

$$\delta \psi^\mu = -i\rho^\alpha \partial_\alpha X^\mu \varepsilon$$

We can derive these transformations by calculating δY^μ explicitly. Using $Q_A = (\partial/\partial\bar{\theta}_A) - i(\rho^\alpha \theta)_A \partial_\alpha$, we have

$$
\begin{aligned}
\delta Y^\mu(\sigma^\alpha, \theta) &= \bar{\varepsilon}Q_A Y^\mu \\
&= \bar{\varepsilon}\left[\frac{\partial}{\partial\bar{\theta}_A} - i(\rho^\alpha \theta)_A \partial_\alpha\right]\left[X^\mu(\sigma^\alpha) + \bar{\theta}\psi^\mu(\sigma^\alpha) + \frac{1}{2}\bar{\theta}\theta B^\mu(\sigma^\alpha)\right] \\
&= \bar{\varepsilon}\left[\frac{\partial}{\partial\bar{\theta}_A}X^\mu + \frac{\partial}{\partial\bar{\theta}_A}\bar{\theta}^B \psi^\mu{}_B(\sigma^\alpha) + \frac{\partial}{\partial\bar{\theta}_A}\frac{1}{2}\bar{\theta}^B \theta^B B^\mu(\sigma^\alpha)\right] \\
&\quad - \bar{\varepsilon}\left[i(\rho^\alpha \theta)_A \partial_\alpha X^\mu + i(\rho^\alpha \theta)_A \partial_\alpha \bar{\theta}_B \psi^\mu + i(\rho^\alpha \theta)_A \partial_\alpha \frac{1}{2}\bar{\theta}\theta B^\mu(\sigma^\alpha)\right] \\
&= \bar{\varepsilon}\left[\psi^\mu(\sigma^\alpha) + \frac{\partial}{\partial\bar{\theta}_A}\frac{1}{2}\bar{\theta}^B \theta^B B^\mu(\sigma^\alpha)\right] \\
&\quad - \bar{\varepsilon}\left[i(\rho^\alpha \theta)_A \partial_\alpha X^\mu + i(\rho^\alpha \theta)_A \partial_\alpha \bar{\theta}_B \psi^\mu\right]
\end{aligned}
$$

To get the last step, we dropped the third-order term which is 0 due to the anticommuting nature of Grassman variables, and we dropped the first term since X^μ does not depend on the supercoordinates. Using the Fierz transformation

$$\theta_A \bar{\theta}_B = -\frac{1}{2}\delta_{AB}\bar{\theta}_C \theta_C$$

We can finally write this as

$$\delta Y^\mu = \bar{\varepsilon}\psi^\mu + \theta(i\bar{\varepsilon}\rho^\alpha \partial_\alpha X^\mu + \bar{\varepsilon}B^\mu) + \theta\bar{\theta}(i\bar{\varepsilon}\rho^\alpha \partial_\alpha \psi^\mu)$$

To write down the SUSY transformations, we simply compare with * and look for corresponding terms in the θ expansion. Doing this we find

$$\delta X^{\mu} = \bar{\varepsilon}\psi^{\mu}$$

$$\delta\psi^{\mu} = -i\rho^{\alpha}\partial_{\alpha}X^{\mu}\varepsilon + B^{\mu}\varepsilon$$

$$\delta B^{\mu} = -i\bar{\varepsilon}\rho^{\alpha}\partial_{\alpha}\psi^{\mu}$$

We can write the action in terms of superfields. Since the superfields are functions of the Grassman variables, we will have to utilize a new kind of integration, called *Grassman integration*. We take a brief detour to describe this now.

Grassman Integration

Another important tool in the supersymmetry toolkit is the technique of Grassman integration. It turns out that the integration of anticommuting Grassman variables is quite a bit different (and actually a lot simpler, although less intuitive) than the ordinary integration of a function of real variables. Even so, we develop the notion of Grassman integration by our wish to preserve one important property of integration. If you integrate a function of a real variable over the entire number line, that integral is translation invariant. That is, let a be some real constant then it must be true that

$$\int_{-\infty}^{\infty} f(x)\,dx = \int_{-\infty}^{\infty} f(x+a)\,dx$$

Now let $\phi(\theta)$ be a function of the Grassman variable θ. We also want integration here to be translation invariant:

$$\int d\theta\, \phi(\theta) = \int d\theta\, \phi(\theta + c)$$

where c is a constant (a Grassman number in this case). To deduce the properties of Grassman integration while preserving translation invariance, we expand the function $\phi(\theta)$ in Taylor:

$$\phi(\theta) = a + b\theta$$

Then,

$$\int d\theta\, \phi(\theta) = \int d\theta\,(a + b\theta)$$

Now, let $\theta \to \theta + c$ and we obtain

$$\int d\theta [a + b(\theta + c)] = \int d\theta(a + b\theta + bc)$$
$$= (a + bc)\int d\theta + b \int d\theta\,\theta$$

In order for this integral to be translation invariant, it cannot depend on c, and so we conclude that

$$\int d\theta = 0$$

when θ is a Grassman variable. By convention, the Grassman integral is normalized to one in the following way:

$$\int d\theta\,\theta = 1$$

So that altogether we have the rule

$$\int d\theta(a + b\theta) = b$$

For double integration over two Grassman coordinates, there is only one rule to remember

$$\int d^2\theta\,\bar{\theta}\,\theta = -2i$$

With these rules in hand, you are on your way to becoming a supersymmetry expert.

A Manifestly Supersymmetric Action

The action written in Chap. 7 [Eq. (7.2)] includes fermionic fields but supersymmetry is not manifest. This situation can be remedied by writing down an action in terms of superfields. The action we use is given by

$$S = \frac{i}{8\pi\alpha'}\int d^2\sigma\,d^2\theta\,\bar{D}Y^\mu\,DY_\mu$$

where,

$$DY^{\mu} = \psi^{\mu} + \theta B^{\mu} - i\rho^{\alpha}\theta \partial_{\alpha} X^{\mu} + \frac{i}{2}\bar{\theta}\theta\rho^{\alpha}\partial_{\alpha}\psi^{\mu}$$

$$\overline{D}Y^{\mu} = \bar{\psi}^{\mu} + B^{\mu}\bar{\theta} + i\bar{\theta}\rho^{\alpha}\, \partial_{\alpha}X^{\mu} - \frac{i}{2}\bar{\theta}\theta\partial_{\alpha}\bar{\psi}^{\mu}\rho^{\alpha}$$

Performing the Grassman integration in the action, we can obtain the component form, which is

$$S = -\frac{1}{4\pi\alpha'}\int d^{2}\sigma(\partial_{\alpha}X_{\mu}\partial^{\alpha}X^{\mu} - i\bar{\psi}^{\mu}\rho^{\alpha}\partial_{\alpha}\psi_{\mu} - B^{\mu}B_{\mu})$$

The equation of motion for B^{μ}, as mentioned earlier, is $B^{\mu} = 0$, which allows us to discard the auxiliary field, and we arrive back at the theory described in Chap. 7.

To see how to arrive at this, you can just apply the rules of Grassman integration, considering θ and $\bar{\theta}$ as separate variables and using $d^{2}\theta = d\theta d\bar{\theta}$. We illustrate by computing a couple of terms. For example,

$$\int d^{2}\theta B^{\mu}\bar{\theta}\,\psi_{\mu} = \int d\theta\int d\bar{\theta}\bar{\theta}B^{\mu}\psi_{\mu}$$

But $\int d\bar{\theta}\bar{\theta} = 1$ and so,

$$\int d\theta\int d\bar{\theta}\bar{\theta}B^{\mu}\psi_{\mu} = \int d\theta B^{\mu}\psi_{\mu} = 0$$

On the other hand,

$$\int d^{2}\theta(B^{\mu}\bar{\theta})(\theta B_{\mu}) = \int d^{2}\theta(\bar{\theta}\theta)B^{\mu}B_{\mu} = -2iB^{\mu}B_{\mu}$$

Using these types of computations, one can transform the manifestly super-symmetric action into the coordinate form to recover the theory of the RNS superstring.

The Green-Schwarz Action

In this section, we use the idea of supersymmetry applied to the space-time coordinates. For worldsheet supersymmetry, we extended the coordinates $\sigma^{\alpha} = (\sigma, \tau)$ of the worldsheet by introducing fermionic super-worldsheet coordinates. Now, we are going to utilize this same idea but apply it to the actual space-time coordinates, which are the bosonic fields $X^\mu(\sigma, \tau)$. This can be done by adding new fields, typically denoted by $\Theta^a(\sigma, \tau)$, which map the worldsheet to fermionic coordinates. Taking the $X^\mu(\sigma, \tau)$ together with the $\Theta^a(\sigma, \tau)$ will enable us to map the worldsheet to superspace. This approach to superstring theory is known as the Green-Schwarz (GS) formalism.

To summarize, when applying worldsheet supersymmetry

- We extend the coordinates (τ, σ) by introducing fermionic coordinates θ^1 and θ^2. This gives us super-worldsheet coordinates.

In this case:

- We are developing an extension of space-time itself, creating a superspace described by the pair $X^\mu(\sigma, \tau)$ and $\Theta^a(\sigma, \tau)$.
- An $N = m$ supersymmetric theory will have $a = 1, ..., m$, or m fermionic coordinates.

A SUPERSYMMETRIC POINT PARTICLE

We introduce the formalism by going back to the simplest case we can describe—a point particle. This will allow us to go over the main ideas without getting bogged down by the formalism. It turns out this approach actually has some direct relevance to string theory anyway. In modern parlance, a point particle is called a *D0-brane*. So the physics we will lay out here is known as the D0-brane action (this is a Dp-brane with $p = 0$). This type of object can be found in the type IIA superstring theory.

The action for a relativistic point particle of mass m can be written as

$$S = \frac{1}{2}\int d\tau \left(\frac{1}{e}\dot{x}^2 - em^2 \right) \tag{9.4}$$

As noted in Chap. 2, e is called the auxiliary field. The action written in this form is well suited to the study of massless particles. Letting $m \to 0$ gives

$$S = \frac{1}{2}\int d\tau \frac{1}{e}\dot{x}^2 \tag{9.5}$$

To make the jump to superspace, we consider the space defined by the pair of coordinates:

$$x^\mu, \theta^{Aa} \tag{9.6}$$

where θ^{Aa} is anticommuting spinor coordinate. In the case we are studying here, for a point particle, these are functions of τ, that is, $\theta^{Aa} = \theta^{Aa}(\tau)$. The index A ranges over the number of supersymmetries in the theory. If there are N of them, then

$$A = 1, \ldots, N$$

Hence, if we have an $N = 2$ supersymmetry, then we have the two fermionic coordinates θ^{1a} and θ^{2a}. You may be a little confused by the notation. We actually have a second index here. The second index is the *spinor index*. Consider a general Dirac spinor. In D dimensions it has $2^{D/2}$ components. So,

$$a = 1, \ldots, 2^{D/2}$$

For Majorana spinors, this number is cut in half. Now, we are actually going to proceed in a manner which is not too different from what you learned for worldsheet supersymmetry. Once again, we consider a constant Majorana spinor that we denote by ε^A (suppressing the spinor index) to emphasize that it is infinitesimal. Now we introduce the following SUSY transformations:

$$\delta x^\mu = i\overline{\varepsilon}^A \Gamma^\mu \theta^A$$
$$\delta \theta^A = \varepsilon^A \tag{9.7}$$
$$\delta \overline{\theta}^A = \overline{\varepsilon}^A$$

In addition, we have to worry about the auxiliary field. We suppose that the SUSY transformation in this case is

$$\delta e = 0 \tag{9.8}$$

The simplest supersymmetric action that can be conceived of is an extension of the action in Eq. (9.5) written as follows:

$$S = \frac{1}{2}\int d\tau \frac{1}{e}(\dot{x}^\mu - i\overline{\theta}^A \Gamma^\mu \dot{\theta}^A)^2 \tag{9.9}$$

Now, since ε^A is a constant, it does not depend on τ and hence $\dot{\varepsilon}^A = 0$. Given that plus Eq. (9.7), it's very easy to see that Eq. (9.9) is invariant under a SUSY transformation. First note that

$$\delta \dot{\theta}^A = \delta\left(\frac{d}{d\tau}\theta^A\right) = \frac{d}{d\tau}(\delta\theta^A) = \frac{d}{d\tau}\varepsilon^A = 0$$

Now of course we can ignore the $1/e$ term when varying the action since the SUSY transformation is Eq. (9.8). Proceeding

$$\delta S = \delta \frac{1}{2} \int d\tau \frac{1}{e} (\dot{x}^\mu - i\overline{\theta}^A \Gamma^\mu \dot{\theta}^A)^2$$

$$= \frac{1}{2} \int d\tau \frac{1}{e} \delta (\dot{x}^\mu - i\overline{\theta}^A \Gamma^\mu \dot{\theta}^A)^2$$

$$= \int d\tau \frac{1}{e} (\dot{x}^\mu - i\overline{\theta}^A \Gamma^\mu \dot{\theta}^A) \delta (\dot{x}^\mu - i\overline{\theta}^A \Gamma^\mu \dot{\theta}^A)$$

$$= \int d\tau \frac{1}{e} (\dot{x}^\mu - i\overline{\theta}^A \Gamma^\mu \dot{\theta}^A) \left[\delta \dot{x}^\mu - i\delta (\overline{\theta}^A \Gamma^\mu \dot{\theta}^A) \right]$$

Ok, now we have

$$\delta(\overline{\theta}^A \Gamma^\mu \dot{\theta}^A) = (\delta\overline{\theta}^A) \Gamma^\mu \dot{\theta}^A + \overline{\theta}^A \Gamma^\mu (\delta\dot{\theta}^A)$$

$$= (\delta\overline{\theta}^A) \Gamma^\mu \dot{\theta}^A$$

$$= \overline{\varepsilon}^A \Gamma^\mu \dot{\theta}^A$$

Using Eq. (9.7) then, we have

$$\delta\dot{x}^\mu - i\delta(\overline{\theta}^A \Gamma^\mu \dot{\theta}^A) = i\overline{\varepsilon}^A \Gamma^\mu \dot{\theta}^A - i\overline{\varepsilon}^A \Gamma^\mu \dot{\theta}^A = 0$$

Therefore, $\delta S = 0$ and the action is invariant under a SUSY transformation.

Since we are dealing with an enlargement of space-time coordinates, take a step back and recall that the actions described in Chap. 2

- Are invariant under space-time translations a^μ.
- Are invariant under Lorentz transformations $\omega^\mu{}_\nu x^\nu$.

We combine these two results in the Poincaré group and note that the action in Eq. (9.4) is invariant under

$$\delta x^\mu = a^\mu + \omega^\mu{}_\nu x^\nu$$

With the enlargement of the space-time coordinates to include the supercoordinates θ^A, and the action in Eq. (9.9), which is invariant under supersymmetry transformations, we now see that we have the *super-Poincaré group*.

In Example 9.1, we illustrate an interesting result. We compute the commutator of two infinitesimal SUSY transformations applied to a space-time coordinate and show that the result is a space-time translation.

EXAMPLE 9.1

Let δ_1 and δ_2 be two infinitesimal supersymmetry transformations on x^μ. Compute $[\delta_1, \delta_2]x^\mu$.

SOLUTION

This is actually rather easy. The commutator is

$$[\delta_1, \delta_2]x^\mu = \delta_1\delta_2 x^\mu - \delta_2\delta_1 x^\mu$$

For the first term, we have

$$\begin{aligned}
\delta_1\delta_2 x^\mu &= \delta_1(i\bar{\varepsilon}_2^A \Gamma^\mu \theta^A)\\
&= i\bar{\varepsilon}_2^A \Gamma^\mu \delta_1\theta^A\\
&= i\bar{\varepsilon}_2^A \Gamma^\mu \varepsilon_1^A
\end{aligned}$$

The second term is

$$\begin{aligned}
\delta_2\delta_1 x^\mu &= \delta_2(i\bar{\varepsilon}_1^A \Gamma^\mu \theta^A)\\
&= i\bar{\varepsilon}_1^A \Gamma^\mu \delta_2\theta^A\\
&= i\bar{\varepsilon}_1^A \Gamma^\mu \varepsilon_2^A
\end{aligned}$$

Therefore

$$[\delta_1, \delta_2]x^\mu = \delta_1\delta_2 x^\mu - \delta_2\delta_1 x^\mu = i\bar{\varepsilon}_2^A \Gamma^\mu \varepsilon_1^A - i\bar{\varepsilon}_1^A \Gamma^\mu \varepsilon_2^A$$

It's a simple exercise to rewrite one term like the other, which gives

$$[\delta_1, \delta_2]x^\mu = -i2\bar{\varepsilon}_1^A \Gamma^\mu \varepsilon_2^A$$

This is just a number, so we can write $[\delta_1, \delta_2]x^\mu = -a^\mu$.

Carrying on, we define a momentum term:

$$\pi_\alpha^\mu = \dot{x}^\mu - i\bar{\theta}^A \Gamma^\mu \partial_\alpha \theta^A \qquad (9.10)$$

where $\alpha = 0, 1, ..., p$ for a general p-brane. In the case of the point particle, which is a 0-brane, only $\alpha = 0$ applies and so

$$\pi_0^\mu = \dot{x}^\mu - i\bar{\theta}^A \Gamma^\mu \dot{\theta}^A \qquad (9.11)$$

In fact, we have already seen that Eq. (9.11) is invariant under a SUSY transformation.

Space-Time Supersymmetry and Strings

We have introduced some of the basic ideas of space-time supersymmetry by considering the point particle (known as the D0-brane). Now we move on to consider the supersymmetric generalization of bosonic string theory. Recall that the action for a bosonic string can be written as

$$S_B = -\frac{1}{2\pi} \int d^2\sigma \sqrt{h} \, h^{\alpha\beta} \partial_\alpha X^\mu \partial_\beta X_\mu \qquad (9.12)$$

Following the procedure used to write down the D0-brane action which was invariant under SUSY transformations, we introduce a new field:

$$\Pi_\alpha^\mu = \partial_\alpha X^\mu - i\bar{\Theta}^A \Gamma^\mu \partial_\alpha \Theta^A \qquad (9.13)$$

This approach is different than the RNS formalism discussed in Chap. 7. The Π_α^μ are actual fermion fields on space-time. In Chap. 7, we had spinors but the ψ^μ were space-time vectors and not genuine fermion fields.

It turns out that in string theory the number of supersymmetries is restricted to $N \leq 2$. If we consider the most general case allowed which has $N = 2$, then there are two fermionic coordinates:

$$\Theta^{1a} \quad \text{and} \quad \Theta^{2a} \qquad (9.14)$$

To get the full action we need to extend Eq. (9.12) in two steps. The first step is to simply add a corresponding piece containing the fermion fields defined in Eq. (9.13). It has basically the same form:

$$S_1 = -\frac{1}{2\pi} \int d^2\sigma \sqrt{h} \, h^{\alpha\beta} \Pi_\alpha^\mu \Pi_{\beta\mu} \qquad (9.15)$$

Now things get hairy for technical reasons. In supersymmetry, there is a local fermionic symmetry called *kappa symmetry*. To avoid getting weighted down with mathematical details in a "Demystified series" book, we are going to leave it to you to read about kappa symmetry in more advanced treatments. Here we simply take it as a given that we need to preserve this kappa symmetry and that we can only do so by adding the following unwieldy piece to the action:

$$S_2 = \frac{1}{\pi} \int d^2\sigma \left[-i\varepsilon^{\alpha\beta} \partial_\alpha X^\mu (\bar{\Theta}^1 \Gamma_\mu \partial_\beta \Theta^1 - \bar{\Theta}^2 \Gamma_\mu \partial_\beta \Theta^2) + \varepsilon^{\alpha\beta} \bar{\Theta}^1 \Gamma^\mu \partial_\alpha \Theta^1 \bar{\Theta}^2 \Gamma_\mu \partial_\beta \Theta^2 \right]$$

$$(9.16)$$

Light-Cone Gauge

As we found in Chap. 7, the quantum theory will force us to take the number of space-time dimensions to be $D = 10$. Since a general Dirac spinor has components $1, ..., 2^{D/2}$, in 10 space-time dimensions a general Dirac spinor is going to have 32 components. I am sure the reader found dealing with 4 components in quantum field theory enough of a headache, what are we going to do with 32 components? Luckily certain restrictions will cut this down dramatically. The first thing to note is that the complete action, which is given by adding up Eqs. (9.12), (9.15), and (9.16)

$$S = S_1 + S_2$$

which is invariant under SUSY transformations and the mysterious local Kappa symmetry only under very specific conditions that restrict the number of space-time dimensions and the type of spinors in the theory. These conditions are given as follows:

- $D = 3$ with Majorana fermions.
- $D = 4$ with Majorana or Weyl fermions.
- $D = 6$ with Weyl fermions.
- $D = 10$ with Majorana-Weyl fermions.

It is clear that we don't live in flatland, so that rules out the first case. The quantum theory forces us to take $D = 10$, which is no surprise since this was explored in Chap. 7. Therefore the spinors that are relevant to our discussion are Majorana-Weyl fermions. This helps us in two ways:

- The Majorana condition makes the spinor components real.
- The Weyl condition eliminates half of the components. This leaves us with a 16-component spinor.

Once again the Kappa symmetry reveals its hand by cutting the number of components by half. So we are left with an eight component Majorana-Weyl spinor.

With this in mind we will proceed with some aspects of light-cone quantization. This procedure imposes several conditions. First let's begin by defining light-cone components of the Dirac matrices. This is done by singling out the $\mu = 9$ component to make the following definitions:

$$\Gamma^+ = \frac{\Gamma^0 + \Gamma^9}{\sqrt{2}} \tag{9.17}$$

$$\Gamma^- = \frac{\Gamma^0 - \Gamma^9}{\sqrt{2}} \tag{9.18}$$

The 10-dimensional Gamma matrices obey the anticommutation relation:

$$\{\Gamma^\mu, \Gamma^\nu\} = -2\eta^{\mu\nu} \qquad (9.19)$$

As a result, notice that

$$(\Gamma^+)^2 = \frac{1}{2}(\Gamma^0 + \Gamma^9)(\Gamma^0 + \Gamma^9)$$

$$= \frac{1}{2}(\Gamma^0\Gamma^0 + \Gamma^9\Gamma^0 + \Gamma^0\Gamma^9 + \Gamma^9\Gamma^9)$$

$$= \frac{1}{2}(\Gamma^0\Gamma^0 + \Gamma^9\Gamma^9 + \{\Gamma^0, \Gamma^9\})$$

$$= \frac{1}{2}(\Gamma^0\Gamma^0 + \Gamma^9\Gamma^9)$$

$$= \frac{1}{2}(-\eta^{00} - \eta^{99}) = \frac{1}{2}(1-1) = 0$$

We summarize this result by saying that Γ^+ and Γ^- are nilpotent, that is,

$$(\Gamma^+)^2 = (\Gamma^-)^2 = 0 \qquad (9.20)$$

To maintain kappa symmetry, a further constraint must be imposed. This is the fact that Γ^+ annihilates the Θ^A :

$$\Gamma^+\Theta^1 = \Gamma^+\Theta^2 = 0 \qquad (9.21)$$

As is usual in the light-cone gauge, we have

$$X^+ = x^+ + p^+\tau \qquad (9.22)$$

It is customary to denote the spinors which contain the remaining eight nonzero components by S^{Aa}. These objects are defined in the following way:

$$\sqrt{p^+}\Theta^1 \to S^{1a}$$

$$\qquad (9.23)$$

$$\sqrt{p^+}\Theta^2 \to S^{2a}$$

Note that there are dotted spinors in the case of type IIA string theory (see *Quantum Field Theory Demystified* if you are not familiar with this).

Making the definition

$$P_\pm^{\alpha\beta} = \frac{1}{2}(h^{\alpha\beta} \pm \varepsilon^{\alpha\beta}/\sqrt{h}) \tag{9.24}$$

we can write down the equations of motion that are derived by adding Eqs. (9.12), (9.15), and (9.16). These are the equations of motion for the GS superstring, which are in general quite complicated:

$$\Pi_\alpha \cdot \Pi_\beta = \frac{1}{2}h_{\alpha\beta}h^{\gamma\delta}\Pi_\gamma \cdot \Pi_\delta \tag{9.25}$$

$$\Gamma \cdot \Pi_\alpha P_-^{\alpha\beta}\partial_\beta\Theta^1 = 0 \tag{9.26}$$

$$\Gamma \cdot \Pi_\alpha P_+^{\alpha\beta}\partial_\beta\Theta^2 = 0 \tag{9.27}$$

Remarkably, in the light-cone gauge, the equations of motion turn out to be very simple. This is because we can simplify the expression:

$$\Pi_\alpha^\mu = \partial_\alpha X^\mu - i\bar{\Theta}^A\Gamma^\mu\partial_\alpha\Theta^A$$

getting the term $\bar{\Theta}^A\Gamma^\mu\partial_\alpha\Theta^A$ to drop out in most cases. Using Eq. (9.21), this is immediate when taking $\mu = +$:

$$\bar{\Theta}^A\Gamma^+\partial_\alpha\Theta^A = 0$$

There is only one nonvanishing term, when $\mu = -$. For the cases where $\mu = i$, we can use the following trick. Consider the fact that

$$\Gamma^+\Gamma^- = \frac{1}{2}(\Gamma^0 + \Gamma^9)(\Gamma^0 - \Gamma^9)$$

$$= \frac{1}{2}(\Gamma^0\Gamma^0 + \Gamma^9\Gamma^0 - \Gamma^0\Gamma^9 - \Gamma^9\Gamma^9)$$

$$= \frac{1}{2}(-\eta^{00} + \eta^{99} + \Gamma^9\Gamma^0 - \Gamma^0\Gamma^9)$$

$$= 1 + \frac{1}{2}(\Gamma^9\Gamma^0 - \Gamma^0\Gamma^9)$$

Doing a similar calculation for $\Gamma^-\Gamma^+$ one can show that the following provides a representation of the identity operator:

$$1 = \frac{\Gamma^+\Gamma^- + \Gamma^-\Gamma^+}{2} \tag{9.28}$$

So, for $\mu = i$, one simply inserts Eq. (9.28) into the term $\bar{\Theta}^A\Gamma^\mu\partial_\alpha\Theta^A$ to make it vanish from the equations of motion. This is only possible in the light-cone gauge, and it can be shown that the equations of motion are

$$\left(\frac{\partial^2}{\partial\sigma^2} - \frac{\partial^2}{\partial\tau^2}\right)X^i = 0 \tag{9.29}$$

$$\left(\frac{\partial}{\partial\tau} + \frac{\partial}{\partial\sigma}\right)S^{1a} = 0 \tag{9.30}$$

$$\left(\frac{\partial}{\partial\tau} - \frac{\partial}{\partial\sigma}\right)S^{2a} = 0 \tag{9.31}$$

Canonical Quantization

Now, we will take a quick look at canonical quantization and the ground state of the type I superstring. First, note that the usual bosonic commutation relations are imposed for the bosonic fields or space-time coordinates $X^\mu(\sigma, \tau)$ and their associated modes. Now we need to extend the theory by defining quantization conditions for the fermionic fields. Since the supercoordinates are fermionic, we apply equal-time anticommutation relations. These are given by

$$\{S^{Aa}(\sigma, \tau), S^{Bb}(\sigma', \tau)\} = \pi\delta^{ab}\delta^{AB}\delta(\sigma - \sigma') \tag{9.32}$$

Open strings in type I theory satisfy boundary conditions given by

$$S^{1a}\Big|_{\sigma=0} = S^{2a}\Big|_{\sigma=0}$$
$$S^{1a}\Big|_{\sigma=\pi} = S^{2a}\Big|_{\sigma=\pi} \tag{9.33}$$

This helps us determine the modal expansions of the fermion fields, which are given by

$$S^{1a} = \frac{1}{\sqrt{2}} \sum_{n=-\infty}^{\infty} S_n^a e^{-in(\tau-\sigma)}$$
$$S^{2a} = \frac{1}{\sqrt{2}} \sum_{n=-\infty}^{\infty} S_n^a e^{-in(\tau+\sigma)} \tag{9.34}$$

The modes satisfy

$$\left\{ S_m^a, S_n^b \right\} = \delta_{m+n,0} \delta^{ab} \tag{9.35}$$

When writing down the spectrum, it is interesting to note that the normal-ordering constant we've had to deal with in previous theories is no longer an issue. This is because it cancels out with the bosonic and fermionic modes (this is supersymmetry after all). The mass-shell condition for the type I open string is

$$\alpha' m^2 = \sum_{n=1}^{\infty} \left(\alpha_{-n}^i \alpha_n^i + n S_{-n}^a S_n^a \right) \tag{9.36}$$

The ground state consists of a bosonic state and its fermionic partner. Both are massless (there is no tachyon state). The bosonic state is a massless vector which is denoted by $|i\rangle$ where $i = 1, \ldots, 8$. The massless fermion partner is given by $|\dot{a}\rangle$ where $\dot{a} = 1, \ldots, 8$ as well. We can transform between the two states as follows:

$$|\dot{a}\rangle = \Gamma^i_{\dot{a}b} S_0^b |i\rangle$$
$$|i\rangle = \Gamma^i_{ab} S_0^b |\dot{a}\rangle \tag{9.37}$$

The states are normalized according to

$$\langle i | j \rangle = \delta_{ij}$$
$$\langle \dot{a} | \dot{b} \rangle = \frac{1}{2} (h\Gamma^+)^{\dot{a}b} \tag{9.38}$$

Summary

In this chapter, we have extended our discussion of supersymmetric string theory. First, we introduced the notion of superfields and showed how to write an action where supersymmetry was manifest. We then discussed some of the central components of space-time supersymmetry, first in the context of the point particle which gives the D0-brane action and then in the case of string theory. The discussion presented in this chapter is far from complete. The reader who is interested in studying superstring theory on a serious level is urged to consult the references.

Quiz

1. Verify that $\theta_A \rightarrow \theta_A + \varepsilon_A$ by calculating $\delta\theta_A = [\bar{\varepsilon}Q, \theta_A]$.
2. Calculate $\{D_A, Q_B\}$.
3. Find $\Gamma^0\Gamma_\mu^\dagger\Gamma^0$.
4. Compute $\{\Gamma_{11}, \Gamma^0\}$.

CHAPTER 10

A Summary of Superstring Theory

Our final foray into the details of superstring theory will be a cursory look at *heterotic superstrings* in the next chapter. The heterotic string theories are a kind of hybrid between the 26-dimensional bosonic theory and the 10-dimensional superstring theory with fermions. This probably sounds intractable, but we will see in a minute how this can possibly work out into a consistent theoretical framework. Before we dive into heterotic strings, let's take a step back and qualitatively summarize the overall picture of string theory.

A Summary of Superstring Theory

Before jumping into some mathematical details about heterotic string theory, let's review the basic structure of string theory. This is a good idea because there are several string theories with different states of the string. Five of the theories are superstring theories, and we also saw that it is possible to construct a theory consisting only of bosons. Actually you can list four different bosonic string theories, which we will do here.

BOSONIC STRING THEORY

We began our look at strings by considering bosonic string theory. This is an unrealistic theory because we know that the real world contains particles that are fermions. Nonetheless bosonic string theory provides an easier framework that can be used to illustrate the key ideas and techniques of string theory.

Some key aspects of bosonic string theory you should remember are

- It introduces the concept of extra spatial dimensions. In order to avoid ghosts (states with negative norm) we were forced to accept that there are 26 space-time dimensions.

- The ground state (the lowest energy or lowest excitation mode of the string) has a negative mass-squared $(m^2 = -1/\alpha')$. This state is called a tachyon. The presence of a tachyon in the theory indicates that the ground state or vacuum is unstable. Note that in relativity, tachyons are particles that travel faster than the speed of light. Therefore the tachyon is a physically unrealistic particle. There is no known way to remove tachyon states from bosonic string theory.

- Bosonic string theory always includes gravity. This is indicated by the presence of a spin-2 state called the graviton. This is a hint that string theories provide a framework for the unification of all known physical interactions.

- Bosonic string theories also include a state called the dilaton. This is a scalar field which is denoted by φ. It is related to the *coupling constant g* via $g = \exp\langle\varphi\rangle$, where $\langle\varphi\rangle$ is the vacuum expectation value of the dilaton field. If you need to brush up on your quantum field theory, note that the coupling constant determines the strength of an interaction. The dilaton field is dynamical (it is space-time dependent), so in string theory we obtain a dramatic result that the string coupling constant can be dynamical. The dilaton is also known as the *gravitational scalar field* and may play a role in the recently discovered nonzero cosmological constant.

Strings can be either open or closed and can be *oriented* or *unoriented*. If a string is oriented, this means that directions along the string are unequivalent. So you can tell which way you're going along the string. By choosing open or closed strings and oriented or unoriented strings, we can actually construct four different bosonic string theories.

If a bosonic string theory has open strings, it automatically includes closed strings as well. This is due to the dynamical behavior of strings. If a string is open, it is possible for the endpoints to join together, forming a closed string state. Let's summarize the four possibilities for bosonic string theory.

If a bosonic string theory only includes closed strings that are oriented, then the spectrum of the theory includes the following states:

- Tachyon
- Massless antisymmetric tensor
- Dilaton
- Graviton

Now, suppose that we only have closed strings, but the theory describes unoriented strings instead. That is, we can't tell which direction we are moving along the string. In this case, the theory no longer includes a massless vector boson. We can summarize the key aspects of the spectrum as

- Tachyon
- Dilaton
- Massless state which is the graviton

Now let's turn to bosonic string theories that include open as well as closed strings. Again, we can choose strings that are oriented and strings that are unoriented. The oriented theory is characterized by

- Tachyon
- Dilaton
- Graviton
- A massless antisymmetric tensor

The closed string and open string tachyons are distinct. Choosing oriented open + closed bosonic string theory gives us

- Tachyon
- Dilaton
- Massless graviton

There is also a massless vector state for open strings, which can be oriented or unoriented. So we see that all bosonic string theories are plagued by the presence of a tachyon state. They have an unstable vacuum and do not include fermions. As a result, we are forced to consider superstring theories.

Superstring Theory

Superstring theory is a generalization of bosonic string theory which extends the theory to include fermions. There are five different superstring theories. We use the word *super* in our description of them because all five theories are based on a theory of physics known as supersymmetry. This theory is characterized by the idea that each fermion has a bosonic partner and vice versa. Some examples are given in Table 10.1.

The existence of supersymmetry is a good indirect test of string theory. For string theory to be true, supersymmetry must exist in nature. At the time of writing no super-partner has ever been discovered, so supersymmetry either doesn't exist in nature or it has been *broken*. One way it could be broken is that the superpartners are extremely massive. This means it would take high energies to see them. The Large Hadron Collider (LHC) set to begin operation in 2008 may be powerful enough to detect supersymmetry.

So, superstring theory includes supersymmetry, which allows us to introduce fermions into the theory. It also includes ghost states, which are removed in an analogous, manner to what we saw in bosonic string theory. When the ghost states are removed we arrive at the second general characteristic of superstring theory:

- There are 10 space-time dimensions.

There are two ways to introduce supersymmetry into string theory, reviewed in Chaps. 7 and 9, respectively:

- The RNS formalism adds supersymmetry to the worldsheet.
- The GS formalism adds supersymmetry to space-time.

We can still characterize superstring theories by noting whether or not they include open and/or closed strings, and whether those strings are oriented or unoriented. In addition, a superstring theory can be characterized by the number of supercharges used in the theory. This is done by saying that a theory with $N = m$

Table 10.1   A listing of some particles and their postulated super-partners.

Partner	Superpartner
Photon (spin 1)	Photino (spin 1/2)
Graviton (spin 2)	Gravitino (spin 3/2)
Quark (spin 1/2)	Squark (spin 0)
Electron (spin 1/2)	Selectron (spin 0)
Gluon (spin 0)	Gluino (spin 1/2)

supercharges has $N = m$ supersymmetry. Finally, we can characterize each superstring theory by the gauge symmetry that it admits. All superstring theories eliminate the tachyon from the spectrum and include a graviton, so superstring theory naturally describes gravity.

TYPE I SUPERSTRING THEORY

Type I superstring theory can be characterized as follows:

- It includes both open and closed strings.
- It describes unoriented strings.
- It has $N = 1$ supersymmetry.
- It has $SO(32)$ gauge symmetry.

In addition, type I superstrings can have charges attached to their ends called *Chan-Paton factors,* a topic we will explore in a later chapter.

TYPE II A

Type II A theory describes closed, oriented superstrings. We can summarize the theory as follows:

- It only includes closed strings.
- It has $N = 2$ supersymmetry.
- It has a $U(1)$ gauge symmetry.

Since this theory only has a $U(1)$ gauge symmetry, it is not large enough to describe all the particle states seen in nature. It can describe gravity and electromagnetism, but cannot describe the weak or strong forces. The theory has two supercharges, and Θ^1 and Θ^2 have opposite chirality. Practically speaking, this means that each fermion has a partner state with opposite chirality.

TYPE II B

Type II B theory also describes closed strings, also oriented. Although it includes fermionic states because it is a superstring theory, it has no gauge symmetry and so can only describe gravity. Like type II A theory, it has $N = 2$ supersymmetry, but Θ^1 and Θ^2 have the same chirality. This remedies the difficulty in type II A theory in that the fermions described in type II B theory do not have partners of opposite chirality. But the lack of a gauge group indicates the theory cannot be the whole story as far as a unified theory of physics.

HETEROTIC *SO*(32)

There are two heterotic theories that both describe closed, oriented strings. A heterotic theory is a kind of fusion between bosonic and superstring theory. The left movers and right movers are treated using different theories. We describe modes moving in one direction using bosonic string theory, and describe the modes moving in the opposite direction using $N = 1$ supersymmetry. The extra 16 dimensions of the bosonic theory are regarded as abstract, mathematical entities rather than actual space-time coordinates (like superspace). There are two heterotic theories, both with large gauge groups that can describe all particles in nature. The first has $SO(32)$.

HETEROTIC $E_8 \times E_8$

Similar to Heterotic $SO(32)$ theory but has the gauge group $E_8 \times E_8$.

Dualities

The state of string theory at this point appears to be a random mess, but the discovery of a set of dualities which relate the five theories amongst themselves saved the day. The fact is that the five theories are all related to one another, and we can transform between them. This has led physicists to believe that there exists an underlying theory. The five superstring theories arise as different aspects or solutions of the underlying theory. While some aspects of the potentially underlying theory have been characterized, the actual underlying theory remains unknown. It goes by the name of *M-theory*.

T-DUALITY

We have already studied one duality in detail in Chap. 8, *T-duality*. To review T-duality relates a theory with a small compact dimension to a theory where that same dimension is large. T-duality relates string theories as follows:

- It relates type II A and type II B theory.
- It relates the two heterotic theories.

T-duality can be summarized by saying that if we transform from a small to a large distance scale we exchange momentum and winding modes (and vice versa). T-duality relates type II A and type II B theory in that if we move from small to large distance in type II A theory, the theory is transformed into type II B theory and vice versa (or switch momentum and winding modes). The same holds for the two heterotic theories. This means that type II A and type II B are really the same theory, and the two heterotic theories are really the same theory.

S-DUALITY

The second big duality that has been discovered is *S-duality*. Remember that a coupling constant determines the strength of an interaction, and in string theory the dilaton field determines the value of the coupling constant. String theories have different coupling constants that are weak or strong. By letting $\varphi \to -\varphi$ where φ is the dilaton field, since the coupling constant is defined from $g = \exp\langle\varphi\rangle$, we see that we can transform a large coupling constant into a small one and vice versa, changing a strong interaction into a weak one and vice versa. This is what S-duality is about. S-duality brings type I superstring theory into the fold. That is, under S-duality

- Type I superstring theory is related to heterotic $SO(32)$ superstring theory.
- Type II B is S-dual to itself.

So, a strong interaction in Type I superstring theory is the same as a weak interaction in heterotic $SO(32)$ theory, and vice versa. In other words, the two theories are really the same theory at different coupling strengths.

Quiz

1. T-duality relates string theories in which way?

    (a) It relates strong interactions in type II A theory to weak interactions in type II B theory.

    (b) It relates strong interactions in heterotic theory to weak interactions in type I theory.

    (c) It relates large and small distance scales and momentum and winding modes in two different theories.

    (d) It only relates large and small distance scales.

2. The most significant difference between superstring theory and bosonic theory is

    (a) Bosonic theory has 16 extra space-time dimensions.

    (b) Superstring theory eliminates tachyon states and incorporates fermions into the theory.

(c) Bosonic string theory eliminates tachyons.

(d) Superstring theory has an unstable vacuum.

3. The difference between types II A and type II B theory is

   (a) Type II A theory is nonchiral and type II B theory is chiral.

   (b) Type II A theory only describes open strings.

   (c) Type II B theory only describes bosons.

   (d) Type II B theory is nonchiral and type II A theory is chiral.

4. The dilaton field

   (a) Is only found in bosonic string theory.

   (b) Is related to the coupling constant, but only in heterotic theory.

   (c) Is related to the coupling constant in all string theories.

   (d) Is known to be a mathematical trick not found in nature.

5. Physicists were excited by dualities because

   (a) They add fermions to the theory.

   (b) They show that the five superstring theories are related, so are different aspects of an underlying, unknown theory.

   (c) They show that bosonic and superstring theories are related, so are different aspects of an underlying, unknown theory.

6. The number of space-time dimensions in string theory

   (a) Is fixed at 26 by an ad hoc assumption.

   (b) Is fixed at 26 for superstring theories and 10 for bosonic string theory, because this eliminates ghost states from the theory.

   (c) Is fixed at 10 for superstring theories and 26 for bosonic string theory, because this eliminates tachyon states from the theory.

   (d) Is fixed at 10 for superstring theories and 26 for bosonic string theory, because this eliminates ghost states from the theory.

CHAPTER 11

Type II String Theories

In this chapter, we will review the states of type II A and type II B superstrings. We will do so using the worldsheet supersymmetry plus Gliozzi-Scherk-Olive (GSO) projection approach because it's a bit simpler, so we will review some of the discussion of Chap. 7. In the next chapter, we will briefly discuss heterotic superstrings.

The *R* and *NS* Sectors

To introduce worldsheet supersymmetry we began with the action [Eq. (7.2)]:

$$S = -\frac{T}{2} \int d^2\sigma \, (\partial_\alpha X^\mu \partial^\alpha X_\mu - i\bar{\psi}^\mu \rho^\alpha \partial_\alpha \psi_\mu)$$

In the light-cone gauge, the fermionic part of the action assumes the form

$$S_F = iT \int d^2\sigma (\psi_- \cdot \partial_+ \psi_- + \psi_+ \cdot \partial_- \psi_+)$$

In type II string theories, we only consider closed strings. Therefore, we apply periodic boundary conditions. There are actually two possibilities. Periodic boundary conditions are known as Ramond (R) boundary conditions:

$$\psi_A^\mu(\sigma) = \psi_A^\mu(\sigma + 2\pi) \tag{11.1}$$

Antiperiodic boundary conditions are called Neveu-Schwarz (NS):

$$\psi_A^\mu(\sigma) = -\psi_A^\mu(\sigma + 2\pi) \tag{11.2}$$

Remember that a closed string has independent left-moving and right-moving modes. We can apply either periodic or antiperiodic boundary conditions to the left- and right-moving modes independently, which will give us four possibilities, as discussed in Chap. 7.

THE R SECTOR

Now we review the R sector in more detail. The left ψ_+^μ and right ψ_-^μ modal expansions are

$$\psi_+^\mu(\sigma,\tau) = \sum_{n \in Z} \tilde{d}_n^\mu e^{-2in(\tau+\sigma)}$$

$$\psi_-^\mu(\sigma,\tau) = \sum_{n \in Z} d_n^\mu e^{-2in(\tau-\sigma)} \tag{11.3}$$

Quantization proceeds using the anticommutation relations:

$$\left\{ d_m^\mu, d_n^\nu \right\} = \left\{ \tilde{d}_m^\mu, \tilde{d}_n^\nu \right\} = \eta^{\mu\nu} \delta_{m+n,0} \tag{11.4}$$

These relations are augmented, of course, by the usual bosonic commutation relations. We also have the number operators:

$$N^{(d)} = \sum_{n=1}^{\infty} n d_{-n} \cdot d_n \quad \text{and} \quad \tilde{N}^{(d)} = \sum_{n=1}^{\infty} n \tilde{d}_{-n} \cdot \tilde{d}_n \tag{11.5}$$

The total number operator for the left-moving and right-moving sectors is given by adding the bosonic number operator

$$N^\alpha = \sum_{n=1}^{\infty} \alpha_{-n} \cdot \alpha_n$$

to Eq. (11.5). We obtain

$$N_L = \tilde{N}^\alpha + \tilde{N}^{(d)}$$
$$N_R = N^\alpha + N^{(d)}$$

(11.6)

Taking $n > 0$, we can define creation and annihilation operators as follows:

- d_{-n}^μ acts as a creation operator by adding n to the eigenvalue of $N^{(d)}$.

- d_n^μ acts as an annihilation operator by subtracting n from the eigenvalue of $N^{(d)}$.

The states are constructed in the usual way by using a fock space (or number states). Letting $n > 0$, the ground state is annihilated by the bosonic and fermionic annihilation operators:

$$\alpha_n^\mu |0\rangle_R = d_n^\mu |0\rangle_R = 0$$

(11.7)

for the right-moving modes, and similarly for the left movers. We can construct an arbitrary state by tracing on the ground state multiple times:

$$|n\rangle_R = \prod_i \prod_j (\alpha_{-n_i})^{p_i} (d_{-m_j})^{q_j} |0\rangle_R$$

(11.8)

For states in the right-moving sector, the expression for the left-moving sector is similar.

Now consider the special case of d_0^μ. The anticommutation relation is

$$\{d_0^\mu, d_0^\nu\} = \eta^{\mu\nu}$$

(11.9)

This is almost the same as the commutation relation obeyed by the Gamma matrices (i.e., the "Dirac algebra"):

$$\{\Gamma^\mu, \Gamma^\nu\} = -2\eta^{\mu\nu}$$

(11.10)

This suggests that these operators are related to the Gamma matrices using

$$\Gamma^\mu = i\sqrt{2}d_0^\mu \tag{11.11}$$

This tells us that the states in the R sector are space-time spinors. We can write the ground state as

$$|0\rangle_R^a$$

where a is a spinor index that ranges over $a = 1, \ldots, 32$. As we have seen earlier, this is because a general Dirac spinor has $2^{D/2}$ components where D is the number of space-time dimensions. Since $D = 10$ for superstring theories, there are 32 components. The state $|0\rangle_R^a$ is a 32-component Majorana spinor.

Now recall that the chirality operator $\Gamma_{11} = \Gamma_0 \Gamma_1 \ldots \Gamma_9$ acts on states $|0\rangle_R^\pm$ of definite chirality according to

$$\Gamma_{11}|0\rangle_R^+ = +|0\rangle_R^+$$
$$\Gamma_{11}|0\rangle_R^- = -|0\rangle_R^- \tag{11.12}$$

States with definite chirality are Majorana-Weyl spinors, which have half the number of components, (16 in this case). We can write the state $|0\rangle_R^a$ as a direct sum of positive and negative chirality states:

$$|0\rangle_R^a = |0\rangle_R^+ \oplus |0\rangle_R^- \tag{11.13}$$

This gives the state [Eq. (11.13)] $16 \oplus 16 = 32$ components. The states $|0\rangle_R^\pm$ are space-time fermions. However, they have the bizarre property that $|0\rangle_R^+$ is bosonic and $|0\rangle_R^-$ is fermionic on the worldsheet.

THE NS SECTOR

In the NS sector, we have the modal expansions of the left- and right-moving fermionic states given by

$$\psi_+^\mu(\sigma,\tau) = \sum_{r \in Z+1/2} \tilde{b}_r^\mu e^{-2ir(\tau+\sigma)}$$
$$\psi_-^\mu(\sigma,\tau) = \sum_{r \in Z+1/2} b_r^\mu e^{-2ir(\tau-\sigma)} \tag{11.14}$$

The expansion coefficients satisfy the anticommutation relations:

$$\left\{ b_r^\mu, b_s^\nu \right\} = \left\{ \tilde{b}_r^\mu, \tilde{b}_s^\nu \right\} = \eta^{\mu\nu} \delta_{r+s,0} \tag{11.15}$$

The number operators are

$$N^{(b)} = \sum_{r=1/2}^{\infty} r b_{-r} \cdot b_r \qquad \tilde{N}^{(b)} = \sum_{r=1/2}^{\infty} r \tilde{b}_{-r} \cdot \tilde{b}_r \tag{11.16}$$

This allows us to define number operators for right- and left-moving modes:

$$\begin{aligned} N_R &= N^\alpha + N^{(b)} \\ N_L &= \tilde{N}^\alpha + \tilde{N}^{(b)} \end{aligned} \tag{11.17}$$

Letting $n > 0$:

- b_{-r}^μ acts as a creation operator, increasing the eigenvalue of $N^{(b)}$ by r.
- b_r^μ acts as an annihilation operator, decreasing the eigenvalue of $N^{(b)}$ by r.

Again we construct the states using a fock space. The ground state is annihilated in the following way:

$$\alpha_n^\mu \left| 0 \right\rangle_{NS} = b_r^\mu \left| 0 \right\rangle_{NS} = 0 \tag{11.18}$$

An arbitrary right-moving state is given by acting multiple times on the ground state as

$$\left| n \right\rangle_{NS} = \prod_i \prod_j (\alpha_{-n_i})^{p_i} (b_{-r_j})^{q_j} \left| 0 \right\rangle_{NS} \tag{11.19}$$

The NS sector results in states that are space-time bosons.

To get the most general space-time state, one forms tensor products of the left-moving sector and the right-moving sector. Each sector can be an NS state or an R state independently of the other, so we have four possible states overall. If both sectors are $\left| NS \right\rangle$ states, then the state of the string is

$$\left| \psi \right\rangle = \left| NS \right\rangle_{\text{left}} \otimes \left| NS \right\rangle_{\text{right}}$$

This state is a space-time boson. If both states are $|R\rangle$ states

$$|\psi\rangle = |R\rangle_{\text{left}} \otimes |R\rangle_{\text{right}}$$

This is also a space-time boson (this is a state called a *bispinor*, since it is constructed out of two spinors).

As described qualitatively in Chap. 7, we can also have left-moving and right-moving modes from different sectors. Of course, there are two possibilities:

$$|\psi\rangle = |R\rangle_{\text{left}} \otimes |NS\rangle_{\text{right}} \quad \text{or} \quad |\psi\rangle = |NS\rangle_{\text{left}} \otimes |R\rangle_{\text{right}}$$

These states are space-time fermions.

The Spin Field

In supersymmetry, one can move between fermionic and bosonic states using a supercharge operator. We now construct such an operator that allows us to move from an NS (bosonic) to an R (fermionic) state.

Let ϕ^k where $k = 1,...,5$ be complex bosonic fields. We can define the spinors ψ^μ in terms of these bosonic fields using a process called *bosonization*. Remember from Chap.7 that the ψ^μ are space-time vectors, so have 10 components (each of which is a spinor). These are defined in terms of the bosonic fields as follows:

$$e^{\pm i\phi^1} = \psi^1 \pm i\psi^2$$
$$e^{\pm i\phi^2} = \psi^3 \pm i\psi^4$$
$$e^{\pm i\phi^3} = \psi^5 \pm i\psi^6$$
$$e^{\pm i\phi^4} = \psi^7 \pm i\psi^8$$
$$e^{\pm i\phi^5} = \psi^9 \pm i\psi^{10}$$

We can construct a spin field operator by taking the product of these quantities, that is

$$S^a = e^{\pm i\phi^{1/2}} \cdots e^{\pm i\phi^{5/2}} \tag{11.20}$$

This is a 32-component $SO(10)$ spinor. It acts like a supercharge, taking the bosonic state $|0\rangle_{NS}$ to the fermionic state $|0\rangle_R^a$:

$$|0\rangle_R^a = S^a |0\rangle_{NS} \tag{11.21}$$

Now we have everything in place to describe the differences between type II A and type II B string theories.

Type II A String Theory

The key aspect of type II A string theory that should stick in your mind which distinguishes it from type II B string theory is *opposite chirality*. That is, for a state

$$|R_1\rangle \otimes |R_2\rangle$$

$|R_1\rangle$ and $|R_2\rangle$ will have opposite chirality.

First, let's note that the total fock space is constructed as follows. First, we form direct sums:

$$\left(|NS\rangle \oplus |R\rangle \right)_{\text{left}} \qquad \left(|NS\rangle \oplus |R\rangle \right)_{\text{right}}$$

Then we form the tensor product:

$$\left(|NS\rangle \oplus |R\rangle \right)_{\text{left}} \otimes \left(|NS\rangle \oplus |R\rangle \right)_{\text{right}}$$

The physical state space is constructed using GSO projection, which will cure the odd problem of the states differing as being fermions or bosons in space-time and on the worldsheet described earlier, as well as allowing us to remove tachyons from the theory. First, we construct an operator which will count up the number of d_{-n}^μ excitations in a state:

$$F = \sum_{n=1}^\infty d_{-n} \cdot d_n \tag{11.22}$$

It is easy to see that $(-1)^F$ tells us if a state has an even or odd number of d_{-n}^μ excitations. Now suppose that $|\psi\rangle$ is a state with an even number of d_0^μ

oscillators and $|\phi\rangle$ is a state with an odd number of d_0^μ oscillators. It so happens that

$$\Gamma_{11}|\psi\rangle = |\psi\rangle \qquad \Gamma_{11}|\phi\rangle = -|\phi\rangle \tag{11.23}$$

Now define

$$\tilde{\Gamma} = \Gamma_{11}(-1)^F \tag{11.24}$$

Then if

$$\tilde{\Gamma}|\psi\rangle = +|\psi\rangle$$

the state $|\psi\rangle$ has an even number of d_{-n}^μ excitations. On the other hand if

$$\tilde{\Gamma}|\psi\rangle = -|\psi\rangle$$

the state $|\psi\rangle$ has an odd number of d_{-n}^μ excitations.

Type II A theory is characterized by states with opposite chiralities. So, the GSO projection of the right movers has the opposite sign of the GSO projection of the left movers. This means that the chirality of space-time fermions will turn out to be opposite. Specifically note that

$$\left(\Gamma_{11}|0\rangle_R\right)_{\text{left}} = -\left(\Gamma_{11}|0\rangle_R\right)_{\text{right}} \tag{11.25}$$

in type II A theory. Now consider the spin field S^a. Since $\Gamma_{11}|0\rangle_R = \pm|0\rangle_R$ and $|0\rangle_R = S^a|0\rangle_{NS}$ where $|0\rangle_{NS}$ is bosonic, we can consider the action of Γ_{11} on the spin field itself. In type II A theory

$$\Gamma_{11}S^a = S^a \qquad \text{and} \qquad \Gamma_{11}\bar{S}^a = -\bar{S}^a \tag{11.26}$$

The action of GSO projection is to take the 32-component spinor into a 16-component Majorana-Weyl spinor.

Type II B Theory

We have seen that type II A theory is characterized by opposite chirality. Type II B theory is characterized by the same chirality. That is, for a state

$$\left|R_1\right\rangle \otimes \left|R_2\right\rangle$$

$\left|R_1\right\rangle$ and $\left|R_2\right\rangle$ will have the same chirality. The left and right movers will have the same GSO projection:

$$\left(\Gamma_{11}\left|0\right\rangle_R\right)_{\text{left}} = \left(\Gamma_{11}\left|0\right\rangle_R\right)_{\text{right}} \tag{11.27}$$

The spinor field satisfies

$$\Gamma_{11}S^a = S^a \qquad \text{and} \qquad \Gamma_{11}\bar{S}^a = \bar{S}^a \tag{11.28}$$

The states

$$\left|0\right\rangle_R^{\text{left}} \otimes b_{-1/2}^\mu \left|0\right\rangle_{NS}^{\text{right}} \qquad \text{and} \qquad b_{-1/2}^\mu \left|0\right\rangle_{NS}^{\text{left}} \otimes \left|0\right\rangle_R^{\text{right}}$$

also have the same chirality.

The Massless Spectrum of Different Sectors

We conclude the chapter by noting the spectrum of states seen in the different sectors, and describing how these are different in type II A and type II B string theories. For those with a background in differential geometry or general relativity, forms are odd in type II A theory and even in type II B theory.

THE $\left|NS\right\rangle \oplus \left|NS\right\rangle$ SECTOR

Type II A and type II B string theories have the same NS-NS sector. The spectrum of states in the NS-NS sector is

- A scalar field φ which is the dilaton described in the last chapter (1 state)
- An antisymmetric gauge field with 28 states
- A symmetric traceless rank two tensor $G_{\mu\nu}$ (this is the graviton) with 35 states

THE $|NS\rangle \oplus |R\rangle$, $|R\rangle \oplus |NS\rangle$ SECTOR

The states in these sectors are superpartners of the dilaton and the graviton. Type II A and type II B theories have the same states, but the states $|NS\rangle \otimes |R\rangle, |R\rangle \otimes |NS\rangle$ each have the same chirality in type II B theory and have the opposite chirality in type II A theory. The states are

- The dilatino, the spin-1/2 superpartner of the dilaton, with 28 states.
- The gravitino, the spin-3/2 fermion which is the superpartner of the graviton. For type II A theory the 2 gravitinos have opposite chirality.

THE $|R\rangle \oplus |R\rangle$ SECTOR

These are boson states formed from the tensor product of two Majorana-Weyl spinors. In type II A theory the left and right states have opposite chirality, while in type II B theory they have the same chirality. For type II A theory, there are two states:

- A vector gauge field
- A "3-form" gauge field (see *Relativity Demystified* for a description of forms)

In type II B theory the states are

- A scalar field
- A two-form gauge field
- A four-form gauge field

Summary

In this chapter, we have summarized the states of type II A and type II B string theories, which are based on the R and NS sectors first introduced in Chap. 7. Type II A theory consists of states that are tensor products of states with opposite chiralities, while type II B theory consists of states that are tensor products of states with the same chiralities. This results in different "particle" states. These theories predict the existence of superpartners that have not yet been observed in nature.

Quiz

1. Type II string theories:

    (a) Include only open strings

    (b) Include open and closed strings

    (c) Include only closed strings

    (d) Include open strings only if they are oriented

2. States in the NS sector can be described in space-time as:

    (a) NS sector states must be combined with R sector states to give space-time fermions, otherwise they do not describe states in space-time.

    (b) Space-time bosons.

    (c) Space-time fermions.

    (d) Space-time bosons for oriented strings, space-time fermions for unoriented strings.

3. The spin field operator $S^a = e^{\pm i\phi^1/2} \cdots e^{\pm i\phi^5/2}$ is what type of spinor? How many components does it have in general?

    (a) $SO(10)$, 16 components

    (b) $SO(10)$, 32 components

    (c) $SO(32)$, 10 components

    (d) $SO(10)$, 10 components

4. The superpartner states in the $\left|NS\right\rangle \otimes \left|R\right\rangle, \left|R\right\rangle \otimes \left|NS\right\rangle$ sectors include:

    (a) The dilatino and the gravitino. For type II A theory the 2 gravitinos have opposite chirality.

    (b) The dilatino and the gravitino. For type II A theory the 2 gravitinos have the same chirality.

    (c) The dilatino and the gravitino. For type II A theory the 2 dilatinos have opposite chirality.

    (d) The dilatino and the gravitino. For type II A theory the 2 dilatinos have the same chirality.

5. In type II B theory the GSO projection can be written as:

(a) $\left(\Gamma_{11}|0\rangle_R\right)_{\text{left}} = -\left(\Gamma_{11}|0\rangle_R\right)_{\text{right}}$

(b) $\left(\Gamma_{11}|0\rangle_R\right)_{\text{left}} = \left(\Gamma_{11}|0\rangle_R\right)_{\text{right}} = 0$

(c) $\left(\Gamma_{11}|0\rangle_R\right)_{\text{left}} = \left(\Gamma_{11}|0\rangle_R\right)_{\text{right}}$

(d) $\left(\Gamma_{11}|0\rangle_R\right)_{\text{left}} = e^{-i\langle\Gamma_{11}\rangle}\left(\Gamma_{11}|0\rangle_R\right)_{\text{right}}$

CHAPTER 12

Heterotic String Theory

In Chap. 10, we learned that there are two string theories that treat the left- and right-moving sectors differently. These theories are called *heterotic string theories* for this reason. The modes are treated as follows:

- The left-moving sector is bosonic.
- The right-moving sector is supersymmetric.

This idea sounds nonsensical because we have learned that bosonic string theory lives in a world of 26 space-time dimensions, while superstring theories live in a world of 10 space-time dimensions. The reason for doing such a radical thing is the following. First, we already know that bosonic string theories by themselves are flawed because they do not incorporate fermions. On the other hand, type II superstrings do not incorporate nonabelian gauge symmetries. This means that the standard model cannot be described by superstring theory as formulated with those theories alone. So type II theories seem to leave us with a description of the universe that lacks electroweak theory and QCD, an unacceptable situation since the real world does include these interactions. One way around this problem is to add charges to the ends of the strings, something we will discuss in Chap. 15, but in this chapter we consider a more successful and elegant approach.

The idea of heterotic strings was originally proposed by Gross, Harvey, Martinec, and Rohm. They proposed a theory of closed superstrings with decoupled left- and right-moving modes that preserves the best aspects of both theories, producing a theory which is large enough and sophisticated enough to incorporate the features we know the theory must have to include the standard model. By making the right-moving modes super-symmetric

- We are able to include fermions in the theory.

- We keep tachyons out of the theory, so it has a stable vacuum.

We incorporate nonabelian gauge theory in the left-moving modes. This is done by adding Majorana-Weyl fermions λ^A to the left-moving sector, without adding supersymmetry. We must eliminate the extra 16 dimensions from the 26-dimension contribution of the bosonic theory. This can be understood by discarding the view that the extra 16 dimensions are space-time dimensions. First, note that

- The right-moving modes are supersymmetric. So, there are 10 bosonic fields X^μ among the right-moving modes.

- We keep 10 bosonic fields X^μ from the left-moving modes to match up with the right-moving modes.

Since $26 = 10 + 16$, we need to cancel the remaining 16 contributions from the unwanted X^μ present in the left-moving sector. Since the λ^A are spinors, we need 32 of them to enable the desired calculation, hence we take $A = 1,...,32$. The symmetry group for the λ^A is $SO(32)$ when all of the λ^A have the same boundary condition. This is the $SO(32)$ heterotic theory.

The Action for $SO(32)$ Theory

We can write down the action as follows:

- It will include a bosonic contribution for left- and right-moving modes for 10 dimensional space-time.

- It will include fermionic spinors to add supersymmetry to the 10 dimensional space-time. These will only be rightmovers.

- It will include a contribution from the left-moving λ^A spinors.

The first two pieces are familiar, we use then Majorana-Weyl fermions ψ^μ which are space-time vectors like those used for worldsheet supersymmetry. So, we have

$$S_1 = -\frac{1}{4\pi\alpha'}\int d^2\sigma(\partial_a X^\mu \partial^a X_\mu - 2i\psi_-^\mu \partial_+ \psi_{\mu-}) \qquad (12.1)$$

The right-moving sector must incorporate supersymmetry. This is done in the same way as in Chap. 7, we incorporate the following transformations:

$$\delta X^\mu = i\varepsilon\psi_-^\mu \qquad \text{and} \qquad \delta\psi_-^\mu = \varepsilon\partial_- X^\mu$$

We will add a second action to incorporate the λ^A. This piece is similar to the fermionic piece used in S_1 but now we are considering left-moving modes and we have to include all 32 λ^A. So the action is

$$S_2 = -\frac{1}{4\pi\alpha'}\int d^2\sigma\left(-2i\sum_{A=1}^{32}\lambda_+^A \partial_- \lambda_+^A\right) \qquad (12.2)$$

The total action for the heterotic string is therefore:

$$\begin{aligned}S &= S_1 + S_2 \\ &= -\frac{1}{4\pi\alpha'}\int d^2\sigma\left(\partial_a X^\mu \partial^a X_\mu - 2i\psi_-^\mu \partial_+ \psi_{\mu-} - 2i\sum_{A=1}^{32}\lambda_+^A \partial_- \lambda_+^A\right)\end{aligned} \qquad (12.3)$$

Quantization of $SO(32)$ Theory

In heterotic string theory, we describe two sectors similar to the R and NS sectors we are already familiar with. These are

- The *periodic sector P*
- The *antiperiodic sector A*

We already know how to handle the bosonic modes and right-moving fermionic modes included in the theory. To develop the full theory we need to quantize the λ^A.

There are no surprises here; the techniques are the same as used before. First, we write down a modal expansion, which in the P sector is

$$\lambda^A(\sigma) = \sum_{n=-\infty}^{\infty} \lambda_n^A e^{-2in\sigma} \tag{12.4}$$

These are fermions, so the coefficients of the expansion λ_n^A are required to satisfy the anticommutation relation:

$$\{\lambda_m^A, \lambda_n^B\} = \delta^{AB}\delta_{m+n,0} \tag{12.5}$$

The A sector, like the NS sector we are already familiar with, is similar except we sum over half-integer quantities. We have

$$\lambda^A(\sigma) = \sum_{r\in z+1/2}^{\infty} \lambda_r^A e^{-2ir\sigma} \tag{12.6}$$

With the similar anticommutation relation

$$\{\lambda_r^A, \lambda_s^B\} = \delta^{AB}\delta_{r+s,0} \tag{12.7}$$

The next step is to construct number operators for the left- and right-moving modes, and then to use these to write down the mass spectrum. We can do this using the NS and R sectors along with GSO projection and the super-Virasoro operators as described in Chap. 7. Or, as we learned in Chap. 9, there are actually two ways to describe supersymmetric modes:

- Using worldsheet supersymmetry together with GSO projection—as applied to NS and R sectors
- Using space-time supersymmetry (the GS formalism)

In the first case, for the NS sector we have a number operator for the right-moving modes:

$$N_R = \sum_{n=1}^{\infty} \alpha_{-n} \cdot \alpha_n + \sum_{r=1/2}^{\infty} r b_{-r} \cdot b_r \tag{12.8}$$

Let $|\psi\rangle$ be a physical state. The mass-shell condition for the NS sector is

$$\left(L_0 - \frac{1}{2}\right)|\psi\rangle = 0 \tag{12.9}$$

In addition, we have the constraints

$$G_r|\psi\rangle = L_m|\psi\rangle = 0 \tag{12.10}$$

where $r, M > 0$. The operator L_0 is given by

$$L_0 = \frac{p^2}{8} + N_R - \frac{1}{2} \tag{12.11}$$

Using the Einstein relation $p^2 - m^2 = 0$, we have

$$p^2 = 4 - 8N_R$$

$$\Rightarrow \alpha' m^2 = 4 - 8N_R \tag{12.12}$$

Now let's quickly review the R sector. Once again we have a mass-shell condition, which in this case is

$$L_0|\psi\rangle = 0 \tag{12.13}$$

This is supplemented by

$$F_m|\psi\rangle = L_m|\psi\rangle = 0 \tag{12.14}$$

for $m \geq 0$. The number operator for the right-moving modes in the R sector is

$$N_R = \sum_{n=1}^{\infty} (\alpha_{-n} \cdot \alpha_n + n d_{-n} \cdot d_n) \tag{12.15}$$

The operator L_0 is given by

$$L_0 = \frac{p^2}{8} + N_R \tag{12.16}$$

since the ordering constant is 0 for the R sector. Hence the mass-shell condition for the R sector gives

$$\alpha' m^2 = 8N_R \tag{12.17}$$

In Chap. 9, we learned that there is a simpler, unified description for superstrings using the GS formalism. Since we are dealing with different physics for the left- and right-moving modes, why not take this approach instead of carrying around the NS and R sectors. Using the GS formalism, the number operator for the right-moving modes is given by

$$N_R = \sum_{n=1}^{\infty} \left(\alpha_{-n} \cdot \alpha_n + n S_{-n}^a S_n^a \right) \qquad (12.18)$$

In the GS formalism, the mass-shell condition gives us an expression for the mass that is similar to what we found for the R sector in the RNS formalism:

$$\alpha' m^2 = 8 N_R \qquad (12.19)$$

There will be *two* number operators for the left-moving modes. One for the P sector, and one for the A sector. Given what we learned in Chap. 11, these can be written down immediately for the case of the λ^A:

$$\tilde{N}_L^P = \sum_{n=1}^{\infty} \left(\alpha_{-n} \cdot \alpha_n + n \lambda_{-n}^A \lambda_n^A \right)$$

$$\tilde{N}_L^A = \sum_{n=1}^{\infty} \alpha_{-n} \cdot \alpha_n + \sum_{r=1/2}^{\infty} r \lambda_{-r}^A \lambda_r^A \qquad (12.20)$$

The left-moving modes must also satisfy the Virasoro constraints. In general, the mass-shell condition is

$$(\tilde{L}_0 - \tilde{a}) |\psi\rangle = 0 \qquad (12.21)$$

There is no supersymmetry for the left-moving modes, so the condition [Eq. (12.21)] is only augmented by

$$\tilde{L}_m |\psi\rangle = 0 \qquad (12.22)$$

for $m > 0$. These constraints must be satisfied for the P and A sectors. So we introduce two normal ordering constants which we denote by $\tilde{a}_P$ and $\tilde{a}_A$. Then Eq. (12.21) becomes the two conditions:

$$(\tilde{L}_0 - \tilde{a}_P)|\psi\rangle = 0 \qquad \text{(P sector)}$$
$$(\tilde{L}_0 - \tilde{a}_A)|\psi\rangle = 0 \qquad \text{(A sector)}$$
$$(12.23)$$

The $\tilde{L}_0$ are not identical in these two equations, since we have the two number operators [Eq. (12.20)]. So we have

$$\tilde{L}_0 = \frac{p^2}{8} + \tilde{N}_L^P - \tilde{a}_P \qquad \text{(P sector)}$$
$$\tilde{L}_0 = \frac{p^2}{8} + \tilde{N}_L^A - \tilde{a}_A \qquad \text{(A sector)}$$
$$(12.24)$$

The task now is to determine the values of $\tilde{a}_P$ and $\tilde{a}_A$. It turns out that this can be done readily given what we know about superstrings. That is

- A periodic boson makes a contribution of 1/24 to the normal ordering constant.

- A periodic fermion makes a contribution of $-1/24$ to the normal ordering constant.

The normal ordering constant is actually formed by the sum

$$\tilde{a} = \text{bosonic contribution } + \text{ fermionic contribution}$$

Going to the light-cone gauge, there are eight transverse bosonic components. We also have the 32 fermions λ^A. So, the total normal ordering constant for the P sector is

$$\tilde{a}_P = 8\left(\frac{1}{24}\right) + 32\left(-\frac{1}{24}\right) = -1 \qquad (12.25)$$

Now for the A sector, we need to know that

- An antiperiodic fermion contributes 1/48 to the normal ordering constant.

In the A sector, the bosonic contribution is the same. Hence

$$\tilde{a}_P = 8\left(\frac{1}{24}\right) + 32\left(\frac{1}{48}\right) = 1 \qquad (12.26)$$

You should be aware that the bosonic and fermionic contributions to the normal ordering constants are due to the zero point energy of the bosonic and fermionic modes. Note also that these zero energy contributions are *finite*.

Using Eqs. (12.24), (12.25), and (12.26) and including the relation obtained for the right-moving modes, we obtain the mass formulas for the P and A sectors:

$$\alpha'm^2 = 8N_R = 8\left(\tilde{N}_L^P + 1\right) \quad \text{(P sector)}$$
$$\alpha'm^2 = 8N_R = 8\left(\tilde{N}_L^A - 1\right) \quad \text{(A sector)}$$

$$(12.27)$$

We have made the obvious leap of faith that the mass must be the same for a given string state whether looking at the right-moving or left-moving modes. We immediately notice that

- A state with zero mass has $N_R = 0$.

Put another way, a state with zero mass in heterotic string theory has the right-moving modes in the ground state. In addition, if $m = 0$, then

- For a state in the P sector, $\tilde{N}_L^P = -1$.
- For a state in the A sector, $\tilde{N}_L^A = +1$.

Now, in ordinary quantum mechanics, you learned that a number operator satisfies $N \geq 0$. So, we must reject the first possibility which translates into

- The P sector contains no massless states.

The Spectrum

To describe the spectrum of the theory, we follow the usual procedure of constructing states which are tensor products of left-moving and right-moving modes:

$$|\psi\rangle = |\text{left}\rangle \otimes |\text{right}\rangle \tag{12.28}$$

For the left movers, we just learned that the P sector contributes nothing to a massless state. Since $\tilde{N}_L^A = +1$, this means that the state from the A sector is in the first excited state. There are two possibilities. It can be a bosonic state:

$$|\text{left}\rangle = \tilde{\alpha}_{-1}^j |0\rangle_L \tag{12.29}$$

Or it can be a fermionic state, since we need to consider the λ^A as well:

$$\left|\text{left}\right\rangle = \lambda^A_{-1/2}\lambda^B_{-1/2}\left|0\right\rangle_L \tag{12.30}$$

For right movers, with $N_R = 0$, we can have a bosonic state:

$$\left|\text{right}\right\rangle = \left|i\right\rangle_R \tag{12.31}$$

Or a fermionic state

$$\left|\dot{a}\right\rangle_R \tag{12.32}$$

Now let's consider the case when the left movers are in the bosonic state [Eq. (12.29)]. The *bosonic sector* is given by the tensor product with the bosonic states of the right movers [Eq. (12.31)]:

$$\left|\psi\right\rangle = \tilde{\alpha}^j_{-1}\left|0\right\rangle_L \otimes \left|i\right\rangle_R \tag{12.33}$$

The states [Eq. (12.33)] can be summarized as follows. The "particle" spectrum includes:

- A scalar, the dilaton
- An antisymmetric tensor state given by $\tilde{\alpha}^j_{-1}\left|0\right\rangle_L \otimes \left|i\right\rangle_R - \tilde{\alpha}^i_{-1}\left|0\right\rangle_L \otimes \left|j\right\rangle_R$
- The graviton which is the state $\tilde{\alpha}^j_{-1}\left|0\right\rangle_L \otimes \left|i\right\rangle_R + \tilde{\alpha}^i_{-1}\left|0\right\rangle_L \otimes \left|j\right\rangle_R$

Now let's take a look at the fermionic sector for the massless states. We can get this by pairing up the bosonic states from the left movers with the fermionic states from the right movers. This is going to give us the superpartners of Eq. (12.23). The states can be written as

$$\left|\psi\right\rangle = \tilde{\alpha}^j_{-1}\left|0\right\rangle_L \otimes \left|\dot{a}\right\rangle_R \tag{12.34}$$

The "particle" states here include

- The superpartner of the dilaton, the dilatino
- The superpartner of the graviton, the gravitino

As you might guess from looking at the particle spectrum, supersymmetry is a vital component of string theory. If particle accelerators never find evidence of

superpartners, the status of string theory will be put in doubt. On the other hand, the discovery of superpartners does not prove string theory, but would be a good indication that the theory is on the right track.

Compactification and Quantized Momentum

In this section, for readers who are curious, we briefly describe a different approach developed by Gross, Harvey, Matinec, and Rohm where compactification is used to construct the heterotic string. We follow the description laid out in Kaku (please see References). In the light-cone gauge, the action for the heterotic string can be written as

$$S = -\frac{1}{4\pi\alpha'}\int d^2\sigma\left(\partial_a X^i \partial^a X_i + \sum_{I=i}^{16}\partial_a X^I \partial^a X_I + i\bar{S}\Gamma^-(\partial_\tau + \partial_\sigma)S\right)$$

The approach used here is to compactify the extra bosonic dimensions to generate the group $E_8 \otimes E_8$. The extra 16 dimensions of the bosonic sector are compactified on a lattice. As described in the previous section, the right-moving sector is supersymmetric. The spinors $S^a(\tau - \sigma)$ have 8 components (so $a = 1,...,8$). Remember we are in the light-cone gauge, so only consider transverse components. The index i is used for the space-time components, in the light-cone gauge $i = 1,...,8$ as well. The remaining index I is used to run over the lattice used to compactify the extra 16 dimensions. So it runs over 1 to 16.

The physics is much the same as the previous analysis. Bosonic states $X^i(\tau + \sigma)$ and $X^i(\tau - \sigma)$ are included in the left- and right-moving sectors, respectively. The right-moving sector also includes the fermionic component $S^a(\tau - \sigma)$, while the states $X^I(\tau + \sigma)$ are in the left-moving sector.

The action is invariant under the supersymmetry transformation:

$$\delta X^i = (p^+)^{-1/2}\bar{\varepsilon}\,\Gamma^i S^a$$
$$\delta S^a = i(p^+)^{-1/2}\Gamma_-\Gamma_\mu(\partial_\tau - \partial_\sigma)X^\mu\varepsilon$$

The following constraints are used to keep each component properly locked away as a left mover or a right mover:

$$(\partial_\tau - \partial_\sigma)X^I = 0$$
$$\Gamma^+ S^a = 0$$

The usual formulas for the modal expansions apply

$$X^i(\tau - \sigma) = \frac{x^i}{2} + \frac{p^i}{2}(\tau - \sigma) + \frac{i}{2}\sum_{n=1}^{\infty}\frac{\alpha_n^i}{n}e^{-2in(\tau - \sigma)}$$

$$X^i(\tau + \sigma) = \frac{x^i}{2} + \frac{p^i}{2}(\tau + \sigma) + \frac{i}{2}\sum_{n=1}^{\infty}\frac{\tilde{\alpha}_n^i}{n}e^{-2in(\tau + \sigma)}$$

$$X^I(\tau + \sigma) = \frac{x^I}{2} + \frac{p^I}{2}(\tau + \sigma) + \frac{i}{2}\sum_{n=1}^{\infty}\frac{\tilde{\alpha}_n^I}{n}e^{-2in(\tau + \sigma)}$$

$$S^a(\tau - \sigma) = \sum_{n=-\infty}^{\infty}S_n^a e^{-2in(\tau - \sigma)}$$

Using this approach, we write down two number operators. The number operator for the right-moving sector is

$$N = \sum_{n=1}^{\infty}\left(\alpha_{-n}^i\alpha_n^i + \frac{1}{2}n\bar{S}_{-n}\Gamma^-S_n\right)$$

For the left-moving sector we have

$$\tilde{N} = \sum_{n=1}^{\infty}\left(\alpha_{-n}^i\alpha_n^i + \alpha_{-n}^I\alpha_n^I\right)$$

The mass can be written in terms of the canonical momentum p^I as

$$\frac{1}{4}m^2 = N + \tilde{N} - 1 + \frac{1}{2}\sum_{I=1}^{16}\left(p^I\right)^2$$

Now, let's see how compactification affects these results. It is easiest to understand by stepping back to compactification of one dimension as we did in Chap. 8. Kaku gives a nice example which we restate here. Take a single-dimensional theory and let

$$x = x + 2\pi R$$

Consider a field defined on this space $\phi(x)$. Since the coordinate is periodic, the field must be also

$$\phi(x) = \phi(x + 2\pi R)$$

Now, we know that we can expand $\phi(x)$ in a Fourier series. That is, we can expand it in p where p is the conjugate momentum for the coordinate x. The expansion looks like

$$\phi(x) = \sum_n \phi_n e^{ipx}$$

Now, of course, the exponential function is 2π periodic. Calculating $\phi(x + 2\pi R)$

$$\phi(x + 2\pi R) = \sum_n \phi_n e^{ip(x+2\pi R)} = \sum_n \phi_n e^{ipx} e^{ip(2\pi R)}$$

The presence of the extra term $e^{ip(2\pi R)}$ means that we must take

$$p = \frac{n}{R}$$

So, we relearn an important rule about compactification:

- Compactification quantizes momentum.

For the heterotic string, we compactify each of the extra bosonic coordinates:

$$X^I = X^I + 2\pi L^I$$

Here, L^I represents the lattice spacing. If we span the lattice with basis vectors e_i^I then

$$L^I = \frac{1}{\sqrt{2}} \sum_{i=1}^{16} n_i e_i^I R_i$$

Here, the R_i are the radii of the compactified dimensions. Now, we use the conjugate momenta p^I from the bosonic states, which are compactified as a generator of translations. We take $2p^I$ to be the generator of translations along the lattice in the Ith direction. The periodicity condition $X^I = X^I + 2\pi L^I$ means that

$$e^{i2\pi p^I \cdot L^I} = 1$$

This can only be true if the canonical momentum has an expansion of the form

$$p^I = \sqrt{2} \sum_{i=1}^{16} a_i \frac{e_i^I}{R_i}$$

where the a_i are integer coefficients of the expansion. This has the same form obtained in the one-dimensional case, where we found $p = n/R$. So, to compactify over a lattice you define basis vectors, then sum up over all directions dividing by the radius in each direction.

Summary

In this chapter, we gave an overview of the heterotic string. Using the usual approach to supersymmetry, we wrote down the action, applied anticommutation relations, defined number operators, and applied the mass-shell condition. As a result, we found a theory which contains the dilaton and graviton and their superpartners among the massless states. Further analysis would show that the theory is large enough to include the standard model. We can summarize this as follows:

- The heterotic theory has left-moving and right-moving currents on the string.
- The right-moving currents carry supersymmetry charges and give fermionic states.
- The left-moving currents carry the conserved charges of Yang-Mills theories.

After describing the basic machinery of the heterotic string, we illustrated how compactification can be used to get rid of the unwanted 16 extra dimensions that are added to the theory by including the left-moving bosonic theory.

Quiz

1. How many states are there for the left-moving sector of the form $\tilde{\alpha}_{-1}^j |0\rangle$?
2. How many states in the left-moving states have the form $\tilde{\alpha}_{-1}^I |0\rangle$?

3. Consider the constraint equations in the compactification section. Why is $(\partial_\tau - \partial_\sigma)X^I = 0$ the only constraint on the bosonic modes for the extra 16 dimensions?

4. Given the action

$$S = -\frac{1}{4\pi\alpha'}\int d^2\sigma\left(\partial_a X^i \partial^a X_i + \sum_{I=i}^{16}\partial_a X^I \partial^a X_I + i\bar{S}\Gamma^-(\partial_\tau + \partial_\sigma)S\right),$$ what are the

commutation relations satisfied by the bosonic modes?

5. Consider the description of the heterotic string in the compactification section and calculate $(1+\Gamma_{11})S^a = 0$.

CHAPTER 13

D-Branes

One of the most interesting developments in string theory over the last decade or so was the realization that the theory could incorporate higher-dimensional extended objects—that is, objects beyond one-dimensional strings. When these objects are associated with Dirichlet boundary conditions we call these extended objects *Dp-branes,* where p indicates the number of spatial dimensions it has. The word "brane" comes about by analogy. In our everyday world of three spatial dimensions, we are familiar with the notion of a membrane, which is a two-dimensional surface that can separate two regions. The idea of a Dp-brane is to generalize this concept to consider an extended object of p dimensions.

If the number of spatial dimensions of the D-brane is equal to the total number of spatial dimensions in the entire space-time, we say that we have a *space-filling brane.* There are three pertinent examples we can think of immediately to illustrate space-filling branes:

- If space-time is just the three spatial dimensions and one time dimension we are used to from everyday life and special/general relativity, then a D3-brane would be a space-filling brane.

- In bosonic string theory, there are 26 space-time dimensions. So, a D25-brane would be a space-filling brane.

- In superstring theory, there are 10 space-time dimensions. So, a space-filling brane has 9 spatial dimensions and is called a D9-brane.

Chances are if you're reading this book you've completed calculus so you're familiar with the notion of a hyperplane. When first getting started, the best way to think about a D-brane is

- It is a hyperplane-like object.

- The endpoints of open strings are attached to it.

This is illustrated in Fig. 13.1. Note, however, that not all D-branes are hyperplanes, but this is a good way to visualize them.

The spatial dimensions not associated with the brane are called the *bulk*. The volume of the brane is called the *world-volume*. Note that time flows everywhere, in the bulk and on the D-brane as well.

A model of our universe has been proposed where we live in a D3-brane and the bulk consists of the remaining extra spatial dimensions. Perhaps the most fundamental physical insight that has resulted from the study of D-branes is that

- The interactions of the standard model (electromagnetism, strong, and weak forces) are constrained to the brane.

- Gravity can escape from the brane. Gravitational forces are distributed in the brane and also throughout the higher dimensions. Hence, the strength of gravity is diluted by the higher dimensions. This explains why its strength is so different from that of the other known forces.

For simplicity, we will discuss branes within the context of bosonic string theory.

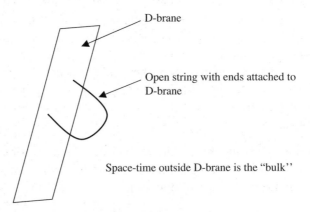

Figure 13.1   A D-brane is a hyperplane-like object to which open strings attach.

The Space-Time Arena

The easiest way to describe a Dp-brane mathematically is to use the light-cone gauge. To specify the D-brane, we need to choose which coordinates will satisfy Neumann boundary conditions and which coordinates satisfy Dirichlet boundary conditions. To use the light-cone gauge, we also need to define light-cone coordinates that will satisfy Neumann boundary conditions, these will include:

- Time
- One spatial coordinate, which we choose to be $X^1(\sigma, \tau)$

For a Dp-brane, we let $i = 2, \ldots, p$ in the light-cone gauge. Then as usual we define:

$$X^{\pm}(\sigma, \tau) = \frac{X^0 \pm X^1}{\sqrt{2}} \tag{13.1}$$

Neumann boundary conditions can be written as

$$\partial_\sigma X^\mu \big|_{\sigma = 0, \pi} = 0 \tag{13.2}$$

So, the coordinates chosen to satisfy Neumann boundary conditions are

$$X^+(\sigma, \tau) \qquad X^-(\sigma, \tau) \qquad X^i(\sigma, \tau) \qquad i = 2, \ldots, p \tag{13.3}$$

Let us suppose that the D-brane is located at $\overline{x}^a$. That is, letting $a = p+1, \ldots, d$:

$$x^a = \overline{x}^a \tag{13.4}$$

The remaining spatial coordinates will satisfy Dirichlet boundary conditions. We use $a = p+1, \ldots, d$ to denote these coordinates. In bosonic string theory we take $d = 25$ while in superstring theory $d = 9$. So the Dirichlet boundary conditions will be applied to

$$X^a(\sigma, \tau) \qquad a = p+1, \ldots, d \tag{13.5}$$

Given $x^a = \bar{x}^a$, the Dirichlet boundary condition can be written as

$$X^a(0,\ \tau) = X^a(\pi,\ \tau) = \bar{x}^a \qquad a = p+1,\ \ldots,\ d \qquad (13.6)$$

Notice that we can also specify the Dirichlet boundary conditions by defining:

$$\delta X^a = X^a(\pi,\ \tau) - X^a(0,\ \tau) \qquad a = p+1,\ \ldots,\ d \qquad (13.7)$$

Then we could write the Dirichlet boundary condition as

$$\delta X^a = 0 \qquad (13.8)$$

The coordinates are divided into two groups and given labels depending on boundary conditions that are applied:

- The coordinates with indices $\mu = \pm$, $i = 2,\ \ldots,\ p$ are called NN coordinates since they satisfy Neumann boundary conditions at both ends.
- The coordinates with indices $a = p+1,\ \ldots,\ d$ are called DD coordinates since they satisfy Dirichlet boundary conditions at both ends.

A simplified illustration of the boundary conditions is shown in Fig. 13.2. To summarize, a Dp-brane is located at $\bar{x}^a$ and has extension along the x^i directions.

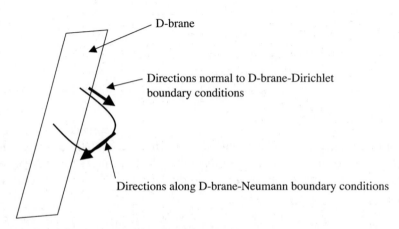

D-brane

Directions normal to D-brane-Dirichlet boundary conditions

Directions along D-brane-Neumann boundary conditions

Figure 13.2   A visualization of the boundary conditions and an open string.

Quantization

Once again we apply the quantization procedure, considering the bosonic string theory case. The first step is to write down the modal expansions. These will be different depending on which coordinates we look at because now we have NN coordinates and DD coordinates. The first step is to write out the modal expansions.

Now let's recall the open string modal expansion, which can be written in the following way:

$$X^{\mu}(\sigma,\tau) = x_0^{\mu} + 2\alpha' p_0^{\mu}\tau + i\sqrt{2\alpha'}\sum_{n\neq 0}\frac{\alpha_n^{\mu}}{n}e^{-in\tau}\cos(n\sigma) \tag{13.9}$$

Taking the derivative of this expression with respect to σ we find

$$\partial_{\sigma}X^{\mu}(\sigma,\tau) = -i\sqrt{2\alpha'}\sum_{n\neq 0}\alpha_n^{\mu}e^{-in\tau}\sin(n\sigma)$$

Clearly this expression satisfies

$$\partial_{\sigma}X^{\mu}\big|_{\sigma=0,\pi} = 0$$

which are the Neumann boundary conditions. So, we take the modal expansion for the X^i to be

$$X^i(\sigma,\tau) = x_0^i + 2\alpha' p_0^i\tau + i\sqrt{2\alpha'}\sum_{n\neq 0}\frac{\alpha_n^i}{n}e^{-in\tau}\cos(n\sigma) \tag{13.10}$$

For the DD coordinates, we really have two requirements that have to be satisfied. We need the summation over the modes $\sum_{n\neq 0}$ in the expansion to vanish at $\sigma = 0,\pi$. This indicates that we should use $\sin(n\sigma)$ instead of $\cos(n\sigma)$ which is in the usual open string expansion. However, we also need $X^a(0,\tau) = X^a(\pi,\tau) = \overline{x}^a$. Looking at the usual modal expansion, this tells us that we must take

$$x_0^a \to \overline{x}^a$$
$$p_0^a \to 0$$

So, the presence of a momentum term in the modal expansion would mean that Dirichlet boundary conditions could not be satisfied. Putting everything together, the modal expansion for the DD coordinates is

$$X^a(\sigma,\tau) = \overline{x}^a + i\sqrt{2\alpha'}\sum_{n\neq0}\frac{\alpha_n^i}{n}e^{-in\tau}\sin(n\sigma) \qquad (13.11)$$

Quantization will involve imposing the usual commutators:

$$\left[X^a(\sigma,\tau),X^b(\sigma',\tau)\right] = i\delta^{ab}\delta(\sigma-\sigma')$$
$$\left[\alpha_m^a,\alpha_n^b\right] = m\delta^{ab}\delta_{m+n,0} \qquad (13.12)$$

Now, for a moment we consider the general light-cone expansion of the string (so for a moment we let $i = 2, ..., 25$). We gauge fix by choosing $\alpha_n^+ = 0$ for all $n \neq 0$ and so

$$X^+(\sigma,\tau) = x_0^+ + 2\alpha'p^+\tau$$

The momentum p^+ is defined as

$$p^+ = \frac{1}{2\alpha'}\alpha_0^+ \qquad (13.13)$$

The light-cone gauge condition is

$$p^+ = \frac{1}{2\alpha'} \qquad (13.14)$$

Now, X^- is an NN coordinate, so

$$X^-(\sigma,\tau) = x^- + 2\alpha'p^-\tau + i\sqrt{2\alpha'}\sum_{n\neq0}\frac{\alpha_n^-}{n}e^{-in\tau}\cos(n\sigma) \qquad (13.15)$$

It follows that

$$\dot{X}^- \pm X'^- = 2\alpha'p^- + \sqrt{2\alpha'}\sum_{n\neq0}\alpha_n^-e^{-in(\tau\pm\sigma)}$$

and

$$(\dot{X}^i \pm X'^i)^2 = \left[2\alpha' p^i + \sqrt{2\alpha'} \sum_{n \neq 0} \alpha_n^i e^{-in(\tau \pm \sigma)} \right]^2$$

Using the light-cone gauge condition, $2\alpha' p^i = 1/p^+$. By writing α_n^- in terms of α_n^i and looking at the $n = 0$ mode, we can use the expression for $(\dot{X}^i \pm X'^i)^2$ to write

$$2\frac{p^-}{p^+} = \frac{p^i p^i}{p^+ p^+} + \frac{1}{\alpha' p^+ p^+} \sum_{n \neq 0} \alpha_{-n}^i \alpha_n^i - \frac{1}{p^+ p^+}$$

So, we find that

$$2 p^+ p^- = \left(p^i p^i + \frac{1}{\alpha'} \sum_{n \neq 0} \alpha_{-n}^i \alpha_n^i - 1 \right) = H \qquad (13.16)$$

where we have introduced the normal-ordering constant $a = 1$ for bosonic string theory into the equations. To transition to the case of a D-brane, all we have to do is have the modes split up into NN and DD coordinates. This means that

$$2 p^+ p^- = \frac{1}{\alpha'} \left[\alpha' p^i p^i + \sum_{n \neq 0} \left(\alpha_{-n}^i \alpha_n^i + \alpha_{-n}^a \alpha_n^a \right) - 1 \right] = H$$

This allows us to write down the mass:

$$m^2 = -p^2 = 2 p^+ p^- - p^i p^i = \frac{1}{\alpha'} \left[\sum_{n \neq 0} \left(\alpha_{-n}^i \alpha_n^i + \alpha_{-n}^a \alpha_n^a \right) - 1 \right] \qquad (13.17)$$

We can define creation and annihilation operators:

$$a^i = \frac{1}{\sqrt{m}} \alpha_m^i \qquad a^{i\dagger} = \frac{1}{\sqrt{m}} \alpha_{-m}^i$$

$$a^a = \frac{1}{\sqrt{m}} \alpha_m^a \qquad a^{i\dagger} = \frac{1}{\sqrt{m}} \alpha_{-m}^a$$

$$\left[a_m^i, a_n^{i\dagger} \right] = \left[a_m^a, a_n^{a\dagger} \right] = \delta_{mn}$$

with all other commutators zero. So the mass can be written in the following way:

$$
m^2 = \frac{1}{\alpha'} \left(\sum_{n=1}^{\infty} \sum_{i=2}^{p} n a_n^{i\dagger} a_n^i + \sum_{m=1}^{\infty} \sum_{a=p+1}^{d} m a_m^{a\dagger} a_m^a - 1 \right)
$$

Due to the presence of the D-brane, the interpretation of the mass has changed. Lorentz invariance is restricted to the brane world-volume, so we view this mass as a mass living in $p+1$ dimensions.

Now, recall that for $a = p+1$, ..., d, the Dirichlet boundary conditions forced us to take the p^a to vanish. This means that states will be of the form

$$
\left| p^+, p^i \right\rangle \tag{13.18}
$$

where $i = 2$, ..., p are the NN coordinates. Since the states only depend on p^i, this means two things:

- Any fields we define are functions of the momenta p^i. String states only have momentum in the NN directions along the brane.

- By writing the Fourier transform, we would see they are functions of the coordinates x^i.

So what does this mean? The fields are defined on coordinates that define the volume of the Dp-brane—they have no coordinate dependence on the $a = p+1$, ..., d and so are zero in the region outside the D-brane. We summarize this by saying that the fields *live on* the Dp-brane.

There are three states we can readily identify. The ground state $\left| p^+, p^i \right\rangle$ is immediately annihilated by the terms a_n^i and a_m^a $\left(\text{that is, } a_n^i \left| p^+, p^i \right\rangle = a_n^a \left| p^+, p^i \right\rangle = 0 \right)$ so the mass is

$$
m^2 \left| p^+, p^i \right\rangle = \frac{1}{\alpha'} \left(\sum_{n=1}^{\infty} \sum_{i=2}^{p} n a_n^{i\dagger} a_n^i + \sum_{m=1}^{\infty} \sum_{a=p+1}^{d} m a_m^{a\dagger} a_m^a - 1 \right) \left| p^+, p^i \right\rangle
$$

$$
= \frac{1}{\alpha'} \left(\sum_{n=1}^{\infty} \sum_{i=2}^{p} n a_n^{i\dagger} a_n^i \left| p^+, p^i \right\rangle + \sum_{m=1}^{\infty} \sum_{a=p+1}^{d} m a_m^{a\dagger} a_m^a \left| p^+, p^i \right\rangle - \left| p^+, p^i \right\rangle \right)
$$

$$
= -\frac{1}{\alpha'} \left| p^+, p^i \right\rangle
$$

Not to be unexpected for the bosonic theory—the ground state is a tachyon. There are two massless states. This is because we have a choice of how we can create the first excited state. We can act on the ground state with $a_1^{i\dagger}$ or with $a_1^{a\dagger}$. Let's consider using $a^{i\dagger}$ first. The state is

$$a_1^{i\dagger}\left|p^+,p^i\right\rangle$$

In this case

$$m^2 a_1^{i\dagger}\left|p^+,p^i\right\rangle = \frac{1}{\alpha'}\left(\sum_{n=1}^{\infty}\sum_{i=2}^{p}na_n^{i\dagger}a_n^i + \sum_{m=1}^{\infty}\sum_{a=p+1}^{d}ma_m^{a\dagger}a_m^a - 1\right)a_1^{i\dagger}\left|p^+,p^i\right\rangle$$

$$= \frac{1}{\alpha'}\left(\sum_{n=1}^{\infty}\sum_{i=2}^{p}na_n^{i\dagger}a_n^i a_1^{i\dagger}\left|p^+,p^i\right\rangle + \sum_{m=1}^{\infty}\sum_{a=p+1}^{d}ma_m^{a\dagger}a_m^a a_1^{i\dagger}\left|p^+,p^i\right\rangle - a_1^{i\dagger}\left|p^+,p^i\right\rangle\right)$$

$$= \frac{1}{\alpha'}\left(\sum_{n=1}^{\infty}\sum_{i=2}^{p}na_n^{i\dagger}a_n^i a_1^{i\dagger}\left|p^+,p^i\right\rangle + \sum_{m=1}^{\infty}\sum_{a=p+1}^{d}ma_m^{a\dagger}a_1^{i\dagger}a_m^a\left|p^+,p^i\right\rangle - a_1^{i\dagger}\left|p^+,p^i\right\rangle\right)$$

Now, $a_m^a\left|p^+,p^i\right\rangle=0$, so

$$m^2 = \frac{1}{\alpha'}\left(\sum_{n=1}^{\infty}\sum_{i=2}^{p}na_n^{i\dagger}a_n^i a_1^{i\dagger}\left|p^+,p^i\right\rangle - a_1^{i\dagger}\left|p^+,p^i\right\rangle\right)$$

Now we use the commutator $\left[a_m^i,a_n^{i\dagger}\right]=\delta_{mn}$ to write $a_n^i a_1^{i\dagger}=\delta_{n1}+a_1^{i\dagger}a_n^i$. And so

$$m^2 = \frac{1}{\alpha'}\left(\sum_{n=1}^{\infty}\sum_{i=2}^{p}na_n^{i\dagger}\left(\delta_{n1}+a_1^{i\dagger}a_n^i\right)\left|p^+,p^i\right\rangle - a_1^{i\dagger}\left|p^+,p^i\right\rangle\right)$$

$$= \frac{1}{\alpha'}\left(\sum_{n=1}^{\infty}\sum_{i=2}^{p}\delta_{n1}na_n^{i\dagger}\left|p^+,p^i\right\rangle - a_1^{i\dagger}\left|p^+,p^i\right\rangle\right)$$

$$= \frac{1}{\alpha'}\left(a_1^{i\dagger}\left|p^+,p^i\right\rangle - a_1^{i\dagger}\left|p^+,p^i\right\rangle\right)=0$$

Hence, the state $a_1^{i\dagger}\left|p^+,p^i\right\rangle$ has mass $m^2=0$. These states are characterized by an index i which denotes coordinates on the brane. Since $i=2, ..., p$, there are a total of $(p+1)-2$ states. Recall that a photon in a $(3+1)$ dimensional theory has two transverse states. So these states are photon states.

The next possibility for a massless state is to act on the ground state with $a_1^{a\dagger}$. You can show that the state $a_1^{a\dagger}|p^+, p^i\rangle$ also has $m^2 = 0$. These states are called *Nambu-Goldstone bosons*. They represent scalar bosons which have arisen from a symmetry breaking of translation invariance in space-time. Excitations of the Nambu-Goldstone bosons $a_1^{a\dagger}|p^+, p^i\rangle$, correspond to displacements of the D-brane in space-time along the coordinate x^a.

The lesson of the string states we have found in the presence of a D-brane is that gauge fields live on the brane.

It turns out that gravity is different. It is not restricted to the brane and can propagate in the bulk.

D-Branes in Superstring Theory

Thinking about superstring theory for a moment on a qualitative level, different types of branes live in different superstring theories. In type IIA theory, only branes with even spatial dimensions are possible. Since $d = 9$ in superstring theory, this means that type IIA superstring theory incorporates branes with the following spatial dimensions:

$$p = 0, 2, 4, 6, 8$$

We met the D0-brane when discussing the supersymmetric point particle in Chap. 9. Now consider type IIB string theory. The dimension of p must be odd, so the theory can contain branes with spatial dimensions:

$$p = -1, 1, 3, 5, 7, 9$$

The case of $p = -1$ might stand out as a little odd. This object is called an *instanton*. It is an object that is forever fixed in time and space-time does not flow for an instanton (thus the name). When $p = 9$ we have a space-filling brane in superstring theory. Note that space-filling branes are possible in type IIB string theory but not in type IIA string theory.

Multiple D-Branes

Having a configuration of multiple D-branes allows for something new—an open string can begin on one brane and end on a different brane. This leads to some interesting results and changes in the mass spectrum. In general we could consider a set of D-branes with spatial dimensions $p, q, r, \ldots$ in various orientations. However, here we will stick to the simplest case, which is to consider two Dp-branes that are parallel but located at different coordinates $\bar{x}_1^a$ and $\bar{x}_2^a$. We will describe this case in a moment and see how the energy from stretching a string between the branes changes the mass spectrum. However, before doing that we take a brief aside to introduce *Chan-Paton factors*.

Chan-Paton factors were introduced into string theory because Yang-Mills theories are necessary to describe the particle interactions of the standard model of particle physics. Before D-branes were known about, the technique used was to attach non-abelian degrees of freedom to the endpoints of open strings. These degrees of freedom were denoted *quark* and *antiquark*, respectively. These names came about by historical accident, string theory was originally proposed as a description of the strong interaction, but it was later displaced from that role by quantum chromodynamics(QCD).

There are $i = 1, \ldots, N$ possible states of a string endpoint. Since an open string has two endpoints, it has two Chan-Paton indices ij. An open string state can be written as:

$$| p; a \rangle = \sum_{i,j=1}^{N} | p; ij \rangle \, \lambda_{ij}^a$$

The λ_{ij}^a are matrices that are called Chan-Paton factors. It turns out that amplitudes obtained when including Chan-Paton factors are invariant under $U(N)$ transformations, which can be transformed into a local $U(N)$ gauge symmetry in space-time. This is exactly what is required for Yang-Mills theories, so it provides a basis for including the standard model in string theory.

After D-branes were discovered, the Chan-Paton indices were reinterpreted. Now we suppose that there are multiple D-branes with integer labels, and string endpoints can be located at D-brane i and j for example. It turns out that multiple D-branes are what give rise to the standard model of particle physics in string theory. In particular, coincident D-branes give rise to massless gauge fields in the following way:

- If there are N coincident Dp-branes, there are N^2 massless gauge fields.
- This characterizes a $U(N)$ Yang-Mills theory on the world-volume of the N coincident D-branes.

We have already seen that a single Dp-brane has a photon state. This is consistent with the outline we are developing here. We have a single D-brane, and the gauge group of the electromagnetic field is $U(1)$. If we add more D-branes in the right way, we can get the number of gauge fields that we want.

As we will see in a moment, strings with endpoints on different branes acquire mass from stretching of the string. Separating coincident D-branes provides a mechanism through which the gauge fields can acquire mass. Now, the gauge group of the electroweak theory is $SU(2)$. There are four gauge fields with quanta:

- The photon
- The W^+ and W^-
- The Z^0

When we have two coincident D-branes, we have $N = 2$ and so there are $N^2 = 4$ gauge fields that transform under $U(2)$. This sounds like the right configuration we need to describe electroweak theory (and you might imagine more branes to include the strong interaction). However, the W^+ and W^- and Z^0 are massive. In quantum field theory, we give them mass using the *Higgs mechanism* (see Chap. 9 and 10 in *Quantum Field Theory Demystified* for a description). In string theory, we separate the two coincident D-branes which will give mass to two of the string states, the states with ends attached to each of the branes. This isn't quite enough since we need one more massive state (and so will need a more complicated D-brane configuration to actually do it right). But you see how the process works.

Now let's quantify the discussion. We consider bosonic string theory again with two D-branes that are parallel. The coordinate locations of the D-branes are given by $\bar{x}_1^a$ and $\bar{x}_2^a$. There are four possibilities for open strings:

- A string has both endpoints attached to D-brane 1.
- A string has both endpoints attached to D-brane 2.
- A string starts on D-brane 1 and ends on D-brane 2.
- A string starts on D-brane 2 and ends on D-brane 1.

Denoting the Chan-Paton indices by (i, j) these possibilities correspond to:

- $(1, 1)$
- $(2, 2)$
- $(1, 2)$
- $(2, 1)$

We already know how the $(1, 1)$ and $(2, 2)$ cases work out—these are open strings with their endpoints attached to the same D-brane. So the spectrum will be unchanged. It includes a tachyon, the photon, and the Nambu-Goldstone boson.

The cases $(1, 2)$ and $(2, 1)$ are string states stretched between the two branes. The descriptions of both cases are the same, so we focus on the $(1, 2)$ case. First, we start with the boundary conditions, which are modified so that the string starts on

D-brane 1 and ends on D-brane 2. Now, let's see how we specify that the string starts on the first D-brane. We quantify this by writing

$$X^a(0,\tau) = \bar{x}_1^a \tag{13.19}$$

To specify that the string ends on the second D-brane, we have:

$$X^a(\pi,\tau) = \bar{x}_2^a \tag{13.20}$$

The oscillator expansions for the NN coordinates are unchanged. However, we need to incorporate the new boundary condition into the oscillator expansion for the DD coordinates. It is now written as:

$$X^a(\sigma,\tau) = \bar{x}_1^a + \frac{1}{\pi}\left(\bar{x}_2^a - \bar{x}_1^a\right)\sigma + i\sqrt{2\alpha'}\sum_{n\neq 0}\frac{\alpha_n^i}{n}e^{-in\tau}\sin(n\sigma) \tag{13.21}$$

It is easy to see that this gives the correct boundary conditions by setting $\sigma = 0, \pi$. You might compare this modal expansion to the modal expansion we got for the DD coordinates earlier, and to the modal expansion for an open string when no D-brane is present. When there is no D-brane, we have a momentum term $p_0^\mu \tau$ which is related to the zeroth mode α_0^μ. In the expansion given here, we have a momentum-like term given by $1/\pi\left(\bar{x}_2^a - \bar{x}_1^a\right)\sigma$. We use this to describe the zeroth mode:

$$\alpha_0^a = \frac{1}{\pi\sqrt{2\alpha'}}\left(\bar{x}_2^a - \bar{x}_1^a\right) \tag{13.22}$$

Notice that this mode does not multiply the timelike coordinate τ, rather it multiplies σ. This tells us that the mode is like a winding mode of the string, but it's really from the stretching of the string from D-brane 1 to D-brane 2. We have to add a term to the expression for the mass to reflect the presence of this additional energy. This is done using:

$$\frac{1}{2\alpha'}\alpha_0^a\alpha_0^a = \left(\frac{\bar{x}_2^a - \bar{x}_1^a}{2\pi\alpha'}\right)^2$$

Previously, with only a single D-brane the mass was given by

$$m^2 = \frac{1}{\alpha'}\left(\sum_{n=1}^{\infty}\sum_{i=2}^{p}na_n^{i\dagger}a_n^i + \sum_{m=1}^{\infty}\sum_{a=p+1}^{d}ma_m^{a\dagger}a_m^a - 1\right)$$

With the extra term due to the stretching, the mass becomes

$$m^2 = \frac{1}{2\alpha'}\alpha_0^a\alpha_0^a + \frac{1}{\alpha'}\left(\sum_{n=1}^{\infty}\sum_{i=2}^{p}na_n^{i\dagger}a_n^i + \sum_{m=1}^{\infty}\sum_{a=p+1}^{d}ma_m^{a\dagger}a_m^a - 1\right) \qquad (13.23)$$

The spectrum of states is modified as follows. Now, the ground state has a mass given by

$$m^2 = -\frac{1}{\alpha'} + \left(\frac{\bar{x}_2^a - \bar{x}_1^a}{2\pi\alpha'}\right)^2$$

The interesting thing about this is that now there are three possibilities for the mass of the ground state which depend on the separation $\bar{x}_2^a - \bar{x}_1^a$ between the two D-branes:

- $\left|\bar{x}_2^a - \bar{x}_1^a\right| < 2\pi\sqrt{\alpha'}$. In this case the mass is negative, so it describes a tachyon state.
- $\left|\bar{x}_2^a - \bar{x}_1^a\right| = 2\pi\sqrt{\alpha'}$. This is a massless state.
- $\left|\bar{x}_2^a - \bar{x}_1^a\right| > 2\pi\sqrt{\alpha'}$. In this case, the ground state is *massive*.

The spectrum also includes one vector and $d - p - 1$ scalars with mass:

$$m^2 = \left(\frac{\bar{x}_2^a - \bar{x}_1^a}{2\pi\alpha'}\right)^2$$

Now let's look at the description in terms of our earlier discussion by considering what happens if the two Dp-branes are coincident. The spectrum then includes:

- Four tachyons
- Four massless vectors
- Four sets of $d - p$ massless scalars

The states transform under 2×2 matrices so the interactions are described by a $U(2)$ gauge theory, which sounds like what we want. Keep in mind that in our simple description given here we are using the bosonic string theory, which is unrealistic and plagued by the tachyon states. But even though it's artificial it gives us an idea of what techniques can be used in the full superstring theory together with more sophisticated D-brane configurations to introduce standard model physics through non-abelian gauge fields living on the brane.

Tachyons and D-Brane Decay

Tachyons can actually describe D-brane decay, so let's say a little bit about that since it shows how they can fit into the overall theory. Consider the action for a scalar field. Suppose that:

$$S = \int d^D x (\partial_\mu \varphi \partial^\mu \varphi + \lambda \varphi^2)$$

Quadratic terms in the potential identify *mass* terms. In the above, we have:

$$\lambda = m^2$$

Now, notice that the quadratic terms indicates a harmonic potential. We can use this to see why the presence of a tachyon indicates an instability of the vacuum. If $m^2 > 0$, then the potential $V(\varphi)$ opens upward, with the minimum located at $\varphi = 0$. On the other hand, if $m^2 < 0$, the parabola opens downward. This means that the point $\varphi = 0$ is unstable. It's like placing a ball at the top of a hill—a small perturbation will cause it to roll down the hill. These potentials are illustrated in Fig. 13.3.

We can expand the potential energy $V(\varphi)$ about its critical points, which tell us where the maxima and minima are, to determine its behavior. To second order it's going to assume the form

$$V(\varphi) = V(\varphi^*) + \lambda (\varphi - \varphi^*)^2 + \cdots$$

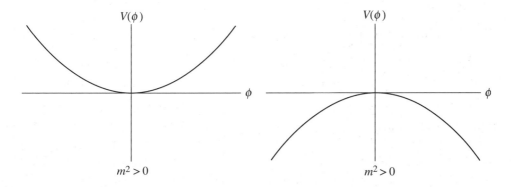

Figure 13.3   A comparison of the potential for $m^2 > 0$ and $m^2 < 0$.

where φ^* is a critical point. The term $\lambda(\varphi-\varphi^*)^2$ is quadratic so is a mass term. In the case of a D-brane, the leading term is given by the tension T. If a tachyon lives on the D-brane, then

$$V(\varphi) = T - \frac{1}{2\alpha'}\varphi^2 + \cdots$$

So, if the potential strays away from $\varphi = 0$ this shows that the D-brane is *losing energy*. What happens is the D-brane decays away into closed string states. Generally speaking this is an artifact of bosonic string theory. In superstring theory there are stable D-brane states. However, in superstring theory you can have an anti-D-brane, which can be coincident with a D-brane. Like particles and antiparticles, they annihilate. This is because there are tachyon states stretched between them.

We consider a simple example. The tachyon potential for a D1-brane coincident with an anti-D1-brane is

$$V(\varphi) = \frac{\lambda}{2}\left(\varphi^2 - \varphi_0^2\right)^2$$

The first step is to find the critical points $V'(\varphi^*) = 0$. The first derivative is

$$V'(\varphi) = 2\lambda\left(\varphi^2 - \varphi_0^2\right)\varphi$$

Setting this equal to 0 we find

$$\varphi^* = 0, \pm\varphi_0$$

The second derivative of the potential is

$$V''(\varphi) = 2\lambda\left(\varphi^2 - \varphi_0^2\right) + 4\lambda\varphi^2$$

Expanding the potential to second order about $\varphi = a$ is

$$V = V(a) + V'(a)(\varphi - a) + \frac{1}{2}V''(a)(\varphi - a)^2 + \cdots$$

We consider the critical point $\varphi^* = 0$ and find

$$V = \frac{\lambda \varphi_0^4}{2} - \lambda \varphi_0^2 \varphi^2 + \cdots$$

The mass term, which multiplies the quadratic term in the expansion is

$$m^2 = -\lambda \varphi_0^2$$

This is a negative mass. So the critical point $\varphi^* = 0$ corresponds to a tachyon.

Summary

Our description of the fascinating topic of D-branes in this chapter barely scratches the surface, but it should help prepare you for more detailed and/or more advanced treatments. We have followed the development by Zweibach in this chapter, so you can see his book for a more detailed analysis, in particular including his discussion of intersecting D6-branes and his accessible discussion of string charge and electromagnetic fields on D-branes (see References). Another good book to consult is the Szabo text listed in the References. For a really detailed (and advanced) description of D-branes, see Clifford V. Johnson's "D-Brane Primer" Johnson's.

Quiz

1. Consider a potential given by $V(\phi) = \frac{A}{6}(\phi^2 - 1)^2$. What are the critical points?

2. Consider $V(\phi) = \frac{A}{6}(\phi^2 - 1)^2$ and expand to second order. Then identify the mass in the case of each critical point found in Prob. 1. Are any of these related to a tachyon?

3. What are the critical points?

4. Which critical point corresponds to a tachyon?

5. If a string is stretched between two parallel D-branes, the ground state acquires mass. Why is that?

CHAPTER 14

Black Holes

Black holes, the collapsed remnants of large stars or the massive central cores of many galaxies, represent an arena where a quantum theory of gravity becomes important. The possibility of the existence of black holes was recognized long ago by the great physicist and mathematician Laplace, but it wasn't until the Schwarzschild solution in general relativity was put forward that these objects and their truly bizarre properties really came into their own. In recent decades the existence of black holes has been established without doubt from observational evidence.

Classically, black holes are remarkably simple objects that can be described by just three properties:

- Mass
- Charge
- Angular momentum

Then Stephen Hawking made a remarkable discovery. In a result that is now famous and is without a doubt known by most readers of this book, Hawking found out that black holes radiate. But this was only the beginning of the story. Black holes have remarkable characteristics that connect them—directly it turns out—to the science of thermodynamics. Black holes have entropy and temperature, and the laws of thermodynamics have analogs that Hawking and his colleagues dubbed the *laws of black hole mechanics*.

One of the most dramatic results of Hawking's work was the implication that black holes are associated with information loss. Physically speaking, we can associate information with pure states in quantum mechanics. In ordinary quantum physics, it is not possible for a pure quantum state to evolve into a mixed state. This is related to the unitary nature of time evolution. What Hawking found was that pure quantum states evolved into mixed states. This is because the character of the radiation emitted by a black hole is thermal, it's purely random—so a pure state that falls into the black hole is emitted as a mixed state. The implication is that perhaps a quantum theory of gravity would drastically alter quantum theory to allow for nonunitary evolution. This is bad because nonunitary transformations do not preserve probabilities. Either black holes destroy quantum mechanics or we have not included an aspect of the analysis that would maintain the missing information required to keep pure states evolving into pure states.

However, it is important to realize that the analysis done by Hawking and others in this context was done using semiclassical methods. That is, a classical space-time background with quantum fields was studied. Given this fact, the results can't necessarily be trusted.

String theory is a fully quantum theory so evolution is unitary. And it turns out that the application of string theory to black hole physics has produced one of the theories most dramatic results to date. Using string theory, it is possible to count the microscopic states of a black hole and compare this to the result obtained using the laws of black hole mechanics (which state that entropy is proportional to area). It is found that there is an exact agreement using the two methods. This is a spectacular result in favor of string theory.

In this chapter we will quickly review the study of black holes in general relativity, state the laws of black hole mechanics, and then illustrate the entropy calculation using string theory.

Black Holes in General Relativity

The existence of black holes is predicted by Einstein's theory of general relativity. Readers interested in a detailed description of black holes in this context may want to consult *Relativity Demystified*.

The Einstein field equations are a set of differential equations which relate the curvature of space-time to the matter-energy content as follows:

$$R_{\mu\nu} - \frac{1}{2} g_{\mu\nu} R + g_{\mu\nu} \Lambda = 8\pi G_4 T_{\mu\nu} \tag{14.1}$$

This equation contains the following elements:

- $R_{\mu\nu}$ is the Ricci tensor. In a moment we will see that it is related to the curvature of space-time through the metric. It can be calculated from the *Riemann curvature tensor* using $R_{\alpha\beta} = R^{\mu}{}_{\alpha\mu\beta}$.

- $g_{\mu\nu}$ is the metric tensor which describes the geometry of space-time.

- R is the Ricci scalar which is computed by contraction of the Ricci tensor.

- Λ is the cosmological constant.

- G_4 is Newton's gravitational constant. It has been noted that this is the gravitational constant in four space-time dimensions, because the form of the gravitational constant depends on the number of space-time dimensions.

- $T_{\mu\nu}$ is the energy-momentum tensor.

The Riemann curvature tensor is

$$R^{\alpha}{}_{\beta\gamma\delta} = \partial_{\gamma}\Gamma^{\alpha}{}_{\beta\delta} - \partial_{\delta}\Gamma^{\alpha}{}_{\beta\delta} + \Gamma^{\varepsilon}{}_{\beta\delta}\Gamma^{\alpha}{}_{\varepsilon\gamma} - \Gamma^{\varepsilon}{}_{\beta\gamma}\Gamma^{\alpha}{}_{\varepsilon\delta} \tag{14.2}$$

where the *Christoffel symbols* are given in terms of the metric tensor as

$$\Gamma_{\alpha\beta\gamma} = \frac{1}{2}(\partial_{\alpha}g_{\beta\gamma} + \partial_{\beta}g_{\gamma\alpha} - \partial_{\gamma}g_{\alpha\beta}) \tag{14.3}$$

When studying the gravitational field outside of the source, the energy-momentum tensor can be set to 0 and we study the *vacuum field equations*. $T_{\mu\nu} = 0$ in a region of empty space-time where no matter or energy is present. The equations are

$$R_{\alpha\beta} - \frac{1}{2}g_{\alpha\beta}R = 0 \tag{14.4}$$

The vacuum field equations describe the structure of space-time outside of a massive body. We use this form of the equations when studying a black hole, because all of the mass is concentrated at a single point at the center called the *singularity*. We can use the vacuum field equations to characterize the structure of the space-time outside this region.

As an aside, note that perturbative string theory adds corrections to the vacuum field equations. These corrections are of the order $O[(\alpha'R)^n]$. If we took the first-order correction from string theory, Eq. (14.4) would be modified as follows:

$$R_{\alpha\beta} - \frac{1}{2}g_{\alpha\beta}R + O(\alpha'R) = 0 \tag{14.5}$$

We will ignore that here, we only mention it for information purposes. Continuing, the simplest case we can imagine is a black hole of mass m which is static (i.e., nonrotating) and spherically symmetric. The metric which describes the space-time outside a black hole of this form is called the *Schwarzschild metric*. The full solution to the vacuum field equations used to arrive at this metric can be found in Chap. 10 of *Relativity Demystified*. We simply state the metric here:

$$ds^2 = -\left(1+\frac{2mG_4}{r}\right)dt^2 +\left(1+\frac{2mG_4}{r}\right)^{-1} dr^2 + r^2 d\Omega^2 \qquad (14.6)$$

where $d\Omega^2 = d\theta^2 + \sin^2\theta\,d\phi^2$. The point $r_H = 2G_4 m$ is called the *horizon*. This appears to be a singular point because setting $r = 2G_4 m$ causes the coefficient of dr to blow up. It can be shown, however, that this is not a real singularity—this singular behavior is just an artifact of the coordinate system. To see this we can calculate a scalar which is an invariant, which gives us insight into the true nature of the horizon. One such invariant is

$$R^{\mu\nu\rho\sigma}R_{\mu\nu\rho\sigma} = \frac{12r_H^2}{r^6} \qquad (14.7)$$

This expression tells us that there is a true singularity at $r = 0$.

Although $r_H = 2G_4 m$ is not a singularity, it is still an important location. This location as we have already indicated denotes the *event horizon*. This is a boundary, in the case of $(3 + 1)$-dimensional space-time the surface of a sphere which divides space-time into the external world and a point of no return. Nothing that crosses the event horizon can ever return to the rest of the universe, not even light. This is why black holes are black, because light cannot escape from inside the horizon.

It will be of interest to study black holes in arbitrary space-time dimension D. With that in mind, before moving on to our next black hole let's define some basic quantities. The first item to note is the volume of a unit sphere in d dimensions. This is given by

$$\Omega_d = \frac{2\pi^{(d+1)/2}}{\Gamma\left(\dfrac{d+1}{2}\right)} \qquad (14.8)$$

where Γ is the gamma factorial function. The radius of the horizon in D-dimensional space-time is given by

$$r_H^{D-3} = \frac{16\pi m G_D}{(D-2)\Omega_{D-2}} \qquad (14.9)$$

Here, G_D is the gravitational constant in D dimensions. In string theory, it depends on the volume of compactified space V, the string-coupling constant g_s, and the string length scale l_s through a 10-dimensional gravitational constant:

$$G_D = \frac{G_{10}}{V} \qquad G_{10} = 8\pi^6 g_s^2 \ell_s^8 \tag{14.10}$$

Now define:

$$h = 1 - \left(\frac{r_H}{r}\right)^{D-3} \tag{14.11}$$

Then, the Schwarzschild metric in D dimensions can be written as

$$ds^2 = -hdt^2 + h^{-1}dr^2 + r^2 d\Omega_{D-2}^2 \tag{14.12}$$

Charged Black Holes

In the introduction we noted that classically a black hole can be completely characterized by its mass, charge, and angular momentum. So in relativity theory there aren't too many choices available to study more complicated black holes. We could have

- A static black hole of mass m (Schwarzschild).
- A static black hole with mass m and electric charge Q.
- A rotating black hole.

A static black hole with electric charge is called a *Reissner-Nordström* black hole, while a rotating black hole is called a *Kerr* black hole. Real astrophysical black holes are best described by Kerr black holes. Stars rotate so when they collapse to a black hole conservation of angular momentum dictates that the black hole will rotate as well. If the rotation is very slow, the Schwarzschild solution would be a good approximation. Real astrophysical black holes, as far as we know, don't carry electrical charge. However, this example is simpler than the Kerr case and it has some advantages which simplify calculations in string theory, so if you're interested in string theory you should familiarize yourself with charged black holes. It is interesting to note that if you add a charge Q to a static black hole in string theory, you can arrive at an exotic black hole that is supersymmetric.

First we define:

$$\Delta = 1 - \frac{2mG_4}{r} + \frac{Q^2 G_4}{r^2} \tag{14.13}$$

Notice that we are basically extending the Schwarzschild solution by adding a Coulomb-type term. The metric for a static, charged black hole is given by

$$ds^2 = -\Delta dt^2 + \Delta^{-1} dr^2 + r^2 d\Omega^2 \tag{14.14}$$

This metric has *two* coordinate singularities which are given by

$$r_\pm = MG_4 \pm \sqrt{(MG_4)^2 - Q^2 G_4} \tag{14.15}$$

The two horizons are denoted by

- r_+ is the *outer horizon*.
- r_- is the *inner horizon*.

The outer horizon is the event horizon—the point of no return when approaching the black hole. Now, before stating our next result, we need to talk a little bit about singularities. Stephen Hawking and Roger Penrose did a great deal of work on singularities in the context of classical general relativity. They found out some interesting results about singularities. If a singularity is present in space-time without a horizon, it is called a *naked singularity*. This is because the horizon, like clothing, keeps you from seeing what's behind the veil. In this case the veil is provided by the fact that light and hence no information can escape from beyond the horizon. The singularity is essentially shut off from the rest of the universe. Hawking and Penrose conjectured that classical physics does not permit the existence of naked singularities.

Charged black holes are related to this concept in the following way. A charged black hole with a mass m is limited in the amount of charge Q that it can carry. It avoids having a naked singularity only if

$$m\sqrt{G_4} \geq |Q| \tag{14.16}$$

When the maximum charge per mass is allowed, we obtain a special type of charged black hole which is called an *extremal black hole*. If you search the literature you will find this term used over and over—extremal black holes are an active area of study. The condition for having an extremal black hole is

$$m\sqrt{G_4} = |Q| \qquad\qquad (14.17)$$

In this case, the radii of the inner and outer horizons are equal. There is only the event horizon whose location is determined to be

$$r_E = r_\pm = mG_4 \qquad\qquad (14.18)$$

Extremal black holes are important because they have unbroken supersymmetry. The metric of an extremal black hole assumes the form

$$ds^2 = -\left(1 - \frac{r_E}{r}\right)^2 dt^2 + \left(1 - \frac{r_E}{r}\right)^{-2} dr^2 + r^2 d\Omega^2 \qquad\qquad (14.19)$$

The key breakthrough for string theory and black holes involved the derivation of black hole entropy from the microscopic states for an extremal black hole in $D = 5$ dimensions. In that case the metric can be written as

$$ds^2 = -\left[1 - \left(\frac{r_E}{r}\right)^2\right]^2 dt^2 + \left[1 - \left(\frac{r_E}{r}\right)^2\right]^{-2} dr^2 + r^2 d\Omega_3{}^2 \qquad\qquad (14.20)$$

For the extremal black hole in five dimensions, the relation between mass and charge becomes

$$m = \frac{Q}{\sqrt{G_5}} = \frac{3\pi r_E^2}{4G_5} \qquad\qquad (14.21)$$

where G_5 is the gravitational constant in five dimensions. The area of the horizon is

$$A = 2\pi^2 r_E^3 \qquad\qquad (14.22)$$

The Laws of Black Hole Mechanics

In the early 1970s, James Bardeen, Brandon Carter, and Stephen Hawking found that there are laws governing black hole mechanics which correspond very closely to the laws of thermodynamics.[1] The zeroth law states that the surface gravity κ at the horizon of a stationary black hole is constant.

The first law relates the mass m, horizon area A, angular momentum J, and charge Q of a black hole as follows:

$$dm = \frac{\kappa}{8\pi} dA + \Omega \, dJ + \Phi \, dQ \tag{14.23}$$

This law is analogous to the law relating energy and entropy. We will see this more precisely in a moment.

The second law of black hole mechanics tells us that the area of the event horizon does not decrease with time. This is quantified by writing:

$$dA \geq 0 \tag{14.24}$$

This is directly analogous to the second law of thermodynamics which tells us that the entropy of a closed system is a nondecreasing function of time. A consequence of Eq. (14.24) is that if black holes of areas A_1 and A_2 coalesce to form a new black hole with area A_3 then the following relationship must hold:

$$A_3 > A_1 + A_2$$

As you probably recall, an analogous relationship holds for entropy. Finally, we arrive at the third law of black hole mechanics. This law states that it is impossible to reduce the surface gravity κ to 0.

The correspondence between the laws of black hole mechanics and thermodynamics is more than analogy. We can go so far as to say that the analogy is taken to be real and exact. That is, the area of the horizon A is the entropy S of the black hole and the surface gravity κ is proportional to the temperature of the black hole. We can express the entropy of the black hole in terms of mass or area. In terms of mass the entropy of a black holes is proportional to the mass of the black

[1] Bardeen, J.M., B. Carter, and S.W. Hawking, The four laws of black hole mechanics, *Comm. Math Phys.* vol. 31, (2), 1973, 161–170.

hole squared. In terms of area, the entropy is 1/4 of the area of the horizon in units of Planck length:

$$S = \frac{A}{4\ell^2_p} \tag{14.25}$$

Computing the Temperature of a Black Hole

Let us compute the temperature in the case of a Schwarzschild black hole. In the Chapter Quiz you will get a chance to try your luck finding the temperature of a charged black hole. We follow a procedure outlined in a note published by P.R. Silva.[2]

We proceed as follows. We perform a Wick rotation $t \rightarrow i\tau$ and write the Schwarzschild metric as

$$ds^2 = -\left(1 - \frac{2G_4M}{r}\right)d\tau^2 + \left(1 - \frac{2G_4M}{r}\right)^{-1}dr^2 + r^2 d\Omega^2 \tag{14.26}$$

Now set

$$Rd\alpha = \left(1 - \frac{2G_4M}{r}\right)^{1/2} d\tau$$

$$dR = \left(1 - \frac{2G_4M}{r}\right)^{-1/2} dr$$

and integrate. We take the limits of integration to be

$$\alpha: \quad 0 \le \alpha \le 2\pi$$
$$\tau: \quad 0 \le \tau \le \beta$$
$$r: \quad 2G_4m \le r' \le r$$

This gives us two relations:

$$2\pi R = (2G_4m)^{-1/2}(r - 2G_4m)^{1/2}\beta \tag{14.27}$$

$$R = 2(2G_4m)^{1/2}(r - 2G_4m)^{1/2} \tag{14.28}$$

---

[2] Available on the arXiv at http://arxiv.org/ftp/gr-qc/papers/0605/0605051.pdf.

Dividing Eq. (14.27) by Eq. (14.28) we obtain

$$2\pi = \frac{\beta}{4G_4 m}$$

$$\Rightarrow \beta = 8\pi G_4 m$$

The β used here is the same one used in thermodynamics, and so we obtain the following expression for the temperature of a Schwarzschild black hole:

$$T = \frac{1}{8\pi m G_4} \tag{14.29}$$

With the temperature in hand, we can proceed ahead to obtain an expression for the entropy. Recalling that the first law of thermodynamics states that $dE = TdS$ (think tedious), using $E = mc^2$ (but taking $c = 1$) we obtain the first law of black hole mechanics for a static, uncharged black hole:

$$dm = Tds \tag{14.30}$$

Hence,

$$mdm = \frac{1}{8\pi G_4} dS$$

Integrating we find

$$\frac{m^2}{2} = \frac{1}{8\pi G_4} S$$

Hence, the entropy of a Schwarzschild black hole is given by

$$S = 4\pi G_4 m^2 \tag{14.31}$$

This confirms our earlier claim, that the entropy is proportional to the mass squared.

Before proceeding, let's quickly refresh our memories. What is entropy anyway? Suppose that we have a density of states $n(E)$ for some microscopic system. The entropy is

$$S = k_B \ln[n(E)] \tag{14.32}$$

where k_B is Boltzman's constant.

Entropy Calculations for Black Holes with String Theory

We will consider two cases, a Schwarzschild black hole where we use a somewhat loose heuristic estimate and a calculation for a five-dimensional charged black hole.

In string theory, the calculation of the entropy of a black hole is easiest in five dimensions. This is due to an amazing property of five-dimensional black holes that arises from supersymmetry. It turns out that supersymmetry allows us to count up string states while taking the string-coupling constant $g_s = 0$, which amounts to considering a set of noninteracting strings—something not really possible in a black hole. This greatly simplifies the calculation and what is remarkable is that the result obtained with no coupling is valid for any coupling strength g_s.

Remember the adiabatic theorem? The procedure used here is something you already know about from ordinary quantum mechanics. In quantum mechanics, you can disturb a system adiabatically so that the energy levels are not disturbed. The adiabatic method is used here—the string-coupling constant is varied adiabatically so that large gravitational forces transition to a weak regime. Entropy, however, is an adiabatic invariant. So while we weaken the string coupling, the entropy remains the same as long as things are done adiabatically.

In string theory, we start with a collection of highly coupled strings and then let the coupling $g_s \to 0$ slowly. We work in the usual $D = 10$ space-time of superstring theory, and need to compactify some extra dimensions to get an effective five-dimensional space-time. Supersymmetry remains unbroken if we compactify dimensions into tiny circles. We compactify the dimensions $x^5, ..., x^9$ leaving us with the remaining five-dimensional space-time described by the coordinates x^0, x^1, x^2, x^3, x^4. The black hole can actually be thought of as two objects—a string carrying a charge Q_1 and a 5-brane with charge Q_2. These charges are winding modes as we will see below.

First we consider an adiabatic process $g_s \to 0$ applied to a Schwarzschild black hole in D dimensions. A straightforward calculation using the laws of black hole mechanics shows that the entropy is given by

$$S = \frac{A}{4G_D} = m_B^{\frac{D-2}{D-3}} G_D^{\frac{1}{D-3}} \Omega_{D-2} \tag{14.33}$$

where m_B is the mass of the black hole. The entropy can be estimated quite simply from string theory considerations. For an excited string, the entropy is proportional to the product of the mass times its length:

$$S \propto m\ell_s \tag{14.34}$$

Now let's use the string theory considerations to find the entropy. This is done by applying the adiabatic procedure to the string coupling $g_s \to 0$ and noting that the entropy remains unchanged during the process. We follow the procedure outlined in Susskind (please see References). What is found is that as $g_s \to 0$, the black hole turns into a single string which has the same entropy as the black hole. The black hole transitions into a single string when the radius of the horizon is the same as the fundamental string length ℓ_s.

The Planck length ℓ_p can be related to the fundamental string length using:

$$\ell_p = g_s^{\frac{2}{D-2}} \ell_s \qquad (14.35)$$

The Schwarzschild radius of the black hole is given in terms of its mass as

$$r_s = (m_B G_D)^{\frac{1}{D-3}} \qquad (14.36)$$

We can approximately take the gravitational constant to be

$$G_D \approx g_s^2 \ell_s^{D-2} \qquad (14.37)$$

So that:

$$\frac{r_s}{\ell_s} = \left(\ell_s m_B g_s^2 \right)^{\frac{1}{D-3}} \qquad (14.38)$$

We are going to be interested in $r_s/\ell_s \to 1$, that is, the point where the Schwarzschild radius approaches the string length. During the adiabatic process, the mass becomes a function of the string-coupling constant, $m = m(g_s)$. If we take the initial coupling to be g_B then $m_B = m(g_B)$. The entropy is a function of the product $m(g_s)\ell_p$, and since the entropy is an adiabatic invariant, this product must be a constant. Using Eq. (14.35) we can write

$$m(g_s) = m_B \left(\frac{g_B^2}{g_s^2} \right)^{\frac{1}{D-2}} \qquad (14.39)$$

Now, $r_s/\ell_s \to 1$ when $m(g_s)\ell_p^{D-2} = \ell_s^{D-3}$ which gives

$$m(g_s)\ell_s = \frac{1}{g_s^2} \qquad (14.40)$$

Recalling that the entropy of an excited string is proportional to the product of its mass and length [Eq. (14.34)], this allows us to write the entropy in terms of the mass of the black hole and the gravitational constant using [Eqs. (14.39) and (14.37)]. This gives a result proportional to Eq. (14.33):

$$S = m_B^{\frac{D-2}{D-3}} G_D^{\frac{1}{D-3}} \propto \frac{A}{4G_D} \tag{14.41}$$

This calculation was not formal by any means. It just relied on some basic considerations from string theory, but it gave the correct result modulo a constant. Now let's turn to the five-dimensional black hole example.

The structure of the five-dimensional geometry is as follows. We take a circular dimension of radius R denoted by S^1 and a four torus T^4 so that

$$T^5 = T^4 \times S^1$$

As mentioned above, the black hole actually has two string components. One is an actual string (a D1-brane) which wraps around S^1 and so has winding modes which will contribute to its mass. The D5-brane wraps around S^1 and also has Kaluza-Klein modes quantized on the circle T^5.

The starting point is the metric given by

$$ds^2 = -\lambda^{-2/3} dt^2 + \lambda^{1/3} \left(dr^2 + r^2 d\Omega_3^2 \right) \tag{14.42}$$

where:

$$\lambda = \prod_{i=1}^{3} \left[1 + \left(\frac{r_i}{r} \right)^2 \right] \tag{14.43}$$

A result from superstring theory that we simply take as a given because it's beyond the scope of our discussion that the BPS condition is satisfied. What this means for us is that charges are additive. The upshot of this is we can write the mass of the black hole as

$$M = M_1 + M_2 + M_3$$

Now the calculation of the entropy is actually quite straightforward. From the metric in Eq. (14.42), there are three radii associated with the horizon. Using Eq. (14.9) with $D = 5$, we can write each of these as

$$r_i^2 = \frac{16\pi m G_5}{3\Omega_3} = \frac{g_s^2 \ell_s^8}{RV} m_i \qquad (14.44)$$

where R is the radius of the circular dimension S^1 and V is the volume of the torus. The individual masses can be calculated from string considerations. The first two masses are due to winding modes. First, the string winding around radius R gives

$$m_1 = \frac{Q_1 R}{g_s \ell_s^2} \qquad (14.45)$$

For the D5-brane, first we have the winding mode which wraps around the circle and torus:

$$m_2 = \frac{Q_5 RV}{g_s \ell_s^6} \qquad (14.46)$$

Then we have a third mass, due to the Kaluza-Klein excitation of the D5-brane along the circular dimension:

$$m_3 = \frac{n}{R} \qquad (14.47)$$

Now let's calculate each of the radii:

$$r_1 = \frac{g_s \ell_s^4}{\sqrt{RV}} \sqrt{m_1} = \frac{\sqrt{g_s} \ell_s^3 \sqrt{Q_1}}{\sqrt{V}} \qquad (14.48)$$

$$r_2 = \frac{g_s \ell_s^4}{\sqrt{RV}} \sqrt{m_2} = \sqrt{g_s} \ell_s \sqrt{Q_5} \qquad (14.49)$$

$$r_3 = \frac{g_s \ell_s^4}{\sqrt{RV}} \sqrt{m_3} = \frac{g_s \ell_s^4}{R\sqrt{V}} \sqrt{n} \qquad (14.50)$$

Now, the area in five dimensions is

$$A = 2\pi^2 r_1 r_2 r_3 = 2\pi^2 \frac{g_s^2 \ell_s^8}{RV} \sqrt{Q_1 Q_5 n}$$

The entropy is

$$S = \frac{A}{4G_5} \tag{14.51}$$

where $G_5 = (2\pi)^5 G_{10}/RV = (2\pi)^5 8\pi^6 g_s^2 \ell_s^8$. Putting everything together we obtain the result:

$$S = 2\pi \sqrt{Q_1 Q_5 n} \tag{14.52}$$

Summary

One of the recent successes of string theory has been its ability to count up the microscopic states of a black hole to calculate its entropy. The result obtained in this manner agrees with the semiclassical expressions, providing strong support for string theory as a quantum theory of gravity.

Quiz

1. Find the temperature of a $D = 4$ charged black hole.

2. The *Hagedorn temperature* is the temperature above which multiple strings would coalesce into a single string. Take the density of string states to be $n = \exp(4\pi m \sqrt{\alpha'})$ and write down the partition function. What condition is necessary for the partition function to be finite? This gives the Hagedorn temperature.

3. Suppose that the sun were to collapse to a black hole. What would be its temperature?

4. Estimate the lifetime of a six solar mass black hole that is evaporating from the Hawking process.

5. Consider a charged, rotating black hole in $D = 5$ dimensions with metric:

$$ds^2 = -\lambda^{-2/3}\left(dt - \frac{a}{r^2}\sin^2\theta d\phi + \frac{a}{r^2}\cos^2\theta d\psi \right)^2 + \lambda^{1/3}\left(dr^2 + r^2 d\Omega_3^2 \right)$$

$$J = \frac{\pi a}{4G_5}$$

By calculating the area of the horizon, estimate the entropy. Take the charges to be the same as for the static charged black hole analyzed in the text.

CHAPTER 15

The Holographic Principle and AdS/CFT Correspondence

In this chapter we will touch on one of the most interesting ideas to come out of the study of quantum gravity and string theory in particular: *the holographic principle*. This is an idea closely related to entropy, so we present it here after we have completed our discussion of black holes and entropy in the last chapter. The holographic principle appears to be a quite general feature of quantum gravity, but we discuss it in the context of string theory. Our discussion largely follows that of Susskind and Witten.[1] Our focus will be on showing how the holographic principle leads to an entropy bound of the type we found for black holes.

---

[1] The topics discussed here are quite advanced, so our discussion will be more qualitative and heuristic. For a detailed exposition you may consult L. Susskind and E. Witten, "The Holographic Bound in Anti-de Sitter Space," http://xxx.lanl.gove/abs/hep-th/9805114; and J. M. Maldacena, "The Large N Limit of Superconformal Field Theories and Super Gravity," *Adv. Theor. Math. Phys.* 2:231–252, 1998.

A Statement of the Holographic Principle

The holographic principle was first proposed by Gerard t'Hooft in 1993 and has been worked on extensively by Leonard Susskind. It can be asserted using two postulates:

- The total information content in a volume of space is equivalent to a theory that lives only on the surface area that encloses the region.
- The boundary of a region of space-time contains at most a single degree of freedom per Planck area.

The holographic principle really applies to gravity and we have already seen it in action when talking about black holes. Information content, which is another way of saying entropy, is about counting the number of states in a system and so is proportional to area. We have already seen that in the case of a black hole that entropy is proportional to the area of the event horizon:

$$S = \frac{A}{4G}$$

where G is Newton's gravitational constant. The area A is measured in Planck units.

This is a surprising result because we would intuitively expect that the number of states is proportional to the *volume* of the enclosed region. Following Susskind, we illustrate that this is in fact the case when gravity is not involved. Imagine that a volume V contains a set of spins on a lattice. We take the lattice spacing to be a, and imagine that the lattice fills the entire volume. Then the total number of spins contained in V is

$$\# \text{ spins } = \frac{V}{a^3}$$

The total number of states the system can have is

$$N = 2^{V/a^3}$$

Using thermodynamics, we arrive at a relationship between the number of states and entropy S:

$$N \propto \exp S$$

Hence we find that $2^{V/a^3} = \exp S$, or taking the logarithm of both sides:

$$S = \ln(2^{V/a^3}) = \frac{V}{a^3} \ln 2$$

We've found what we intuitively expect—the entropy (and by extension the amount of information) in the region is proportional to the volume. After all we started off assuming we had a lattice of spins that filled the volume—so what else could we get?

For black holes we found something very different. In that case, the entropy is directly proportional to the *area* of the even horizon. So in some sense, gravity must be different from other interactions. It turns out that the case of a black hole provides the *maximum entropy* that a gravitational system can have.

A Qualitative Description of AdS/CFT Correspondence

The framework of the holographic principle which comes out of string/M-theory is known as AdS/CFT (anti-de Sitter/conformal field theory) correspondence. We can quantitatively describe the space-time using AdS space in five dimensions. The five-dimensional AdS model has a boundary with four dimensions that looks like flat space with three spatial directions and one time dimension.

The AdS/CFT correspondence involves a duality, something we're already familiar with from our studies of superstring theories. This duality is between two types of theories:

- Five-dimensional gravity
- Super Yang-Mills theory defined on the boundary

By "super" Yang-Mills theory we mean theory of particle interactions with supersymmetry. The holographic principle comes out of the correspondence between these two theories because Yang-Mills theory, which is happening on the boundary, is equivalent to the gravitational physics happening in the five-dimensional AdS geometry. So the Yang-Mills theory can be colloquially thought of as a hologram on the boundary of the real five-dimensional space where the five-dimensional gravitational physics is taking place.

The Holographic Principle and M-Theory

Now let's make our description more quantitative. In the final chapter of the book we discuss stringy cosmology. There we will encounter a model of space-time that has sprung out of string/M-theory that might in fact describe our actual universe. That same model has a nice application in the topic of this chapter as well. The model is a five-dimensional AdS space. It can be described as follows.

We start with a five-dimensional AdS space. In a nutshell, this is a four-dimensional spatial ball and an infinite time axis. The radius of the ball is $0 \le r < 1$. The radius of curvature is denoted by R, and we lump the remaining spatial dimensions together into a unit three-sphere denoted by Ω. The metric which describes the AdS is written as

$$ds^2 = \frac{R^2}{(1-r^2)^2}[(1+r^2)dt^2 - 4dr^2 - 4r^2 d\Omega^2]$$

Note that there are different, equivalent ways to write this metric which you might encounter elsewhere. AdS space has negative curvature and acts like a cavity of size R with reflecting walls. Light or objects and reflect off the boundary and return to the center (see "The Illusion of Gravity" by Juan Maldacena in *Scientific American*, November 2005, for a nice popular level description of AdS).

For us, we are interested in superstring theory. The number of space-time dimensions in superstring theory is $D = 10$. So the complete space is

$$AdS \otimes S^5$$

where S^5 is a unit five-sphere containing the remaining dimensions from string theory. If we denote the extra five coordinates by y_5 they are incorporated into our metric by adding a term Rdy_5^2. We can imagine compactifying these dimensions to a very small size so that they can be effectively ignored. So the universe can be effectively treated as the five-dimensional "bulk" which is the interior of the sphere and the boundary which is the surface. The surface has three spatial dimensions and time.

In the M-theory picture, the world we know is in essence a "shadow" or hologram living on the boundary of a larger dimensional universe. The physics is divided as follows:

- The boundary conformal theory lives on the surface of the sphere at $x = 1$. These are the particles and interactions of the standard model, plus any supersymmetric extension of it.
- Gravity is everywhere.

- But, gravity can propagate into the bulk. In the bulk, which is the interior volume of the AdS sphere, gravity is the only interaction. Inside the ball of the AdS geometry, the theory is *supergravity*. We won't get into supergravity in this book but you can look it up on the arXiv if interested in learning about it.

The conformal theory that describes particles and their interactions is supersymmetric and is called *super Yang-Mills theory* or SYM for short. The gauge group for SYM is $SU(N)$. So the AdS/CFT correspondence can be framed as follows:

- There is a super Yang-Mills theory with $SU(N)$ on the surface of the ball.

- There is bulk supergravity in the interior of the ball.

In string theory, the number of degrees of freedom for the SYM is constrained by three factors:

- The fundamental string length
- The string coupling
- The curvature of AdS space

The number of degrees of freedom for SYM is $\sim N^2$ since the gauge group is $SU(N)$ and it has a gauge coupling g_{YM}. The constraint on N is quantified in the following relationship:

$$R = \ell_s (g_s N)^{1/4}$$

The gauge-coupling is related to the string-coupling constant as:

$$g_{YM}^{\;2} = g_s$$

Now we would like to introduce a cutoff in the bulk. We divide up the sphere into little cells such that the total number of cells in the sphere is $\sim \delta^{-3}$ for some cell δ. That is,

- We cut off the information storage capacity by replacing the continuum of space by cells of size δ.

- There is a single degree of freedom in each cell.

With the total number of degrees of freedom for the SYM theory proportional to N^2, we find that the total number of degrees of freedom with the cutoff is

$$N_{dof} = \frac{N^2}{\delta^3} = A \frac{N^2}{R^3}$$

Now since $R = \ell_s (g_s N)^{1/4}$ we can write

$$N_{dof} = \frac{AR^5}{\ell_s^{\,8} g_s^2}$$

In five-dimensions, the Newton gravitational constant is

$$G_5 = \frac{\ell_s^8 g_s^2}{R^5}$$

Hence we find that

$$N_{dof} = \frac{A}{G_5}$$

This agrees with the holographic principle, and is the same as the result obtained for black holes with the exception of the factor of 1/4.

More Correspondence

In this section we describe connections between the supergravity theory of the bulk and the SYM of the boundary. We can convert between bulk variables and SYM variables as follows. Let E_{SYM} be energy on the boundary and M be the energy in the bulk. They are related as

$$E_{SYM} = RM$$

Temperature is related in the same way:

$$T_{SYM} = RT$$

where T is the temperature in the bulk. Now consider a thermal Yang-Mills state with temperature T_{SYM}. The entropy is

$$S = N^2 (T_{SYM})^3$$

A thermal state of temperature T_{SYM} corresponds to an AdS Schwarzschild black hole at the center of the AdS ball.

Using $T_{SYM} = RT$ and $R = \ell_s (g_s N)^{1/4}$ we obtain

$$(TR)^3 = \frac{S g_s^2 \ell_s^8}{R^8}$$

Now if we take $S = A/4G$ then we find

$$T_{SYM}^3 = \frac{A}{R^3}$$

Now we regulate the SYM so that the maximum T_{SYM} is $1/\delta$. Then we find the maximum area to be

$$A_{max} = \frac{R^3}{\delta^3}$$

Regulation of the super Yang-Mills theory on the boundary gives a holographic description with one bit per Planck area.

An interesting result derived by Susskind and Witten is the *IR-UV connection*. This relates IR divergences in the bulk to UV divergences on the boundary. Consider a string in the bulk that ends on the boundary. The ends of the string correspond to a point charge in the Yang-Mills theory. Now, just thinking back to the self-energy of an electron, you will realize that a point charge in the Yang-Mills theory has a divergent infinite self-energy. This is an UV divergence. The divergence of the bulk string is proportional to $1/\delta$, while δ plays the role of a short distance regulator for UV divergence in SYM theory.

The energy of the string is linearly divergent at the boundaries. Since this divergence is softer, we say that it is an IR divergence. The propagator for a particle of mass m in the bulk is given by

$$\Delta = \frac{\delta^m}{\left| X_1 - X_2 \right|^m}$$

where we have relgulated the area using $A \approx R^3/\delta$ and $\delta \ll 1$. Super Yang-Mills theory is a conformal field theory. Remember Chap. 5? We learned how to calculate operator product expansions there. For super Yang-Mills theory:

$$Y(X_1)Y(X_2) = \mu^{-p}\left| X_1 - X_2 \right|^p$$

You can see that you can transform between these two expressions. What this means is that a propagator for a particle of mass m in the bulk can be transformed into a power law in the conformal field theory on the boundary.

Summary

In this chapter we provided a brief and heuristic introduction to two interesting ideas that have sprung from string theory: the holographic principle and the AdS/CFT correspondence. These two ideas are related. The holographic principle tells us that for an enclosed volume, the informational content of the volume can be described by an equivalent theory that lives on the bounding surface area. This notion is codified in black hole mechanics where the entropy of the black hole is proportional to the area of the horizon, not the volume it encloses. The AdS/CFT correspondence describes a five-dimensional universe where five-dimensional supergravity in the bulk is equivalent to a super Yang-Mills conformal field theory on the boundary.

Quiz

A solution of supergravity gives the metric for a D-brane as:

$$ds^2 = F(z)(dt^2 - dx^2) - F(z)^{-1} dz^2$$

where $F(z) = \left(1 + \dfrac{ag_s N}{z^4}\right)^{-1/2}$.

1. Find an expression for $F(z)$ in the limit $\dfrac{ag_s N}{z^4} \gg 1$.

2. Using your answer to Prob. 1, find a new expression for the metric.

3. The holographic principle can be best described by

   (a) The informational content of a region is encoded in its volume.

   (b) The informational content of a region can be described entirely by the surface area.

   (c) Fields living in the bulk are not equivalent to fields living on the bounding surface.

4. In AdS/CFT correspondence, the number of degrees of freedom available to the super Yang-Mills theory on the boundary is

   (a) Independent of the AdS geometry.

   (b) Related to the string coupling strength only.

   (c) Is related to the string coupling strength and the fundamental string length.

   (d) Is related to the fundamental string length only.

5. In AdS/CFT correspondence, the number of degrees of freedom is

(a) Proportional to the area of the bounding surface and to Newton's gravitational constant.

(b) Proportional to the area of the bounding surface and inversely proportional to Newton's gravitational constant.

(c) Proportional to the area of the bounding surface and the fundamental string length.

(d) Proportional to the area of the bounding surface and the string coupling constant.

CHAPTER 16

String Theory and Cosmology

Conventional cosmology, which grew out of general relativity, astrophysics, and quantum field theory, proposes that the universe began with a "big bang" at a finite time in the past with an inflationary rush, and will expand forever until the universe dies with a whimper, as a result of increasing entropy eventually sapping the useful life out of it. Proposals which originated in string/M-theory have led to different cosmological models. These models have the unexpected and shocking ability to describe the universe before the big bang. Based on a brane-world-type universe, they involve the collision of two branes which get rid of the "singularity" of big-bang theory and replace it with an eternal universe, which could be described as "cyclic." In this chapter, we give an overview of some of the cosmological models that have arisen from string/M-theory. Unfortunately, the details of these models using string/M-theory are well beyond the scope of this book, so our description will be more of a qualitative nature. The motivated reader is urged to consult the references for details. Cosmology is sure to be an active area of research in the coming years with many new and possibly unexpected developments.

Einstein's Equations

In the previous chapter we introduced the Einstein field equations, which give a classical description of gravity. In this chapter, we discuss the application of the Einstein field equations to cosmology. For a detailed description of the study of cosmology in the context of general relativity, please see *Relativity Demystified* and any of the references contained therein.

Cosmology is the study of the evolution of the universe as a whole. The starting point is the Robertson-Walker metric:

$$ds^2 = -dt^2 + a^2(t)d\Sigma^2 \tag{16.1}$$

Here, $d\Sigma^2$ represents the spatial part of the metric. The function $a(t)$ is called the *scale factor*. It characterizes the spatial size of the universe and how it changes with time. The *Hubble constant* is given by

$$H = \frac{\dot{a}}{a} \tag{16.2}$$

We can characterize the spatial structure of the universe by a *curvature constant K*. If the space is flat, has negative curvature (a saddle) or has positive curvature (a sphere), then $K = 0, -1, +1$, respectively. Observational evidence indicates that our universe is flat.

The behavior of the universe with time is determined by starting with a given metric believed to describe the overall structure of the universe, and then using it to work out the components of the curvature tensor. Then we can solve the Einstein field equations either with or without matter present. This can also be done with or without a cosmological constant.

In standard cosmology treatments, space is assumed to be *isotropic*, meaning that it is the same in all directions. We may not want to make that assumption in string theory where some spatial dimensions are treated differently.

There are two cosmological models that come up rather repeatedly. A *de Sitter universe* is one without matter (a vacuum solution of Einstein's field equations), with flat space, and a positive cosmological constant. An *anti-de Sitter universe* (sometimes denoted AdS) is a vacuum solution to the Einstein field equations with positive cosmological constant and negative scalar curvature.

Inflation

The cosmological models studied in relativity theory are only a part of modern cosmology. The second piece which is needed to explain known data is *inflation*. The standard big-bang model begins the universe with a singularity and it expands and cools with its dynamics evolving according to Einstein's equations. Interestingly, the universe exhibits a great deal of uniformity on large scales that the standard big-bang model is hard pressed to explain.

To understand the type of uniformity we are talking about, we can think of everyday life. Imagine heating a cup of tea in the microwave and then taking it out and setting it on the counter. Over time, the cup of tea will cool and if we leave it there long enough, it will reach an equilibrium point where it is the same temperature as its surroundings.

The same kind of behavior has occurred on the largest scales of the universe. If we examine the universe on large scales where we divide it up into cubes that have sides which are on the order of hundreds of millions of light years across, we find

- **Homogeneity:**   On large scales on the average the universe is the same everywhere. That is each cube has the same galaxy density, the same mass density, and the same luminosity.
- **Isotropy:**   We have already mentioned that standard cosmology assumes the universe is isotropic, or the same in every direction. Observation bears this out to an incredibly high degree.

The problem with standard big-bang theory and these observations is that the universe evolved too quickly for equilibrium in the sense we described with the cup of tea, could have occurred. There would not have been enough time for light signals to connect different spatial regions, so how could they have "communicated" so as to end up in exactly the same configuration?

Another problem with standard big-bang cosmology is known as the *flatness problem*. The universe is flat and the mass density of the early universe was apparently so exactly fine-tuned to give the observed flatness that it is hard to imagine how this could be coincidence. The *critical mass density* is defined in terms of the Hubble constant:

$$\rho_c = \frac{3H^2}{8\pi G} \tag{16.3}$$

where G is Newton's gravitational constant. Now define

$$\Omega = \frac{\rho}{\rho_c} \tag{16.4}$$

where ρ is the actual mass density in the universe. Now let Λ be the cosmological constant. If

$$\Omega + \frac{\Lambda}{3H^2} \tag{16.5}$$

exceeds 1, then the universe is a closed space like a sphere. If it is less than one, then it is an open space with negative curvature (like a saddle). If it is *exactly* 1, then the universe is a flat open space. Observation indicates that the universe is a flat open space, so the flatness problem boils down to the equation of why early conditions in the universe fixed the mass density so close to the critical mass density.

The issues which cannot be explained by classical physics-homogeneity, isotropy, and the flatness problem can be explained by a theory known as *inflation*. This is a theory that proposes that the early universe went through a brief phase of exponential expansion. Just prior to the phase of exponential expansion, all regions of the universe were causally connected. This explains the homogeneity and isotropic problems. The expansion is driven by a scalar field ϕ (a quantum field called the inflaton) which has negative pressure. This acts like a repulsive gravitational field causing different regions of the universe to repel one another and to expand outward.

The inflaton field is believed to have a false vacuum, which is a metastable point that is higher in energy than the true vacuum (the lowest energy state). For a brief period, the inflaton was at the false vacuum and could cause inflation, then it "rolled down the hill" to the true vacuum or lowest energy state. During the expansion, the total energy of the universe remains constant (as it must). During inflation, the energy of matter, which is positive, is increasing exponentially. Energy from the inflaton field can be used to actually create matter through Einstein's equation $E = mc^2$.

As matter is added to the universe, the gravitational field gets larger as well. The gravitational field has negative energy density. So the increasing negative energy of the gravitational field balances out the increasing positive energy of matter keeping the total energy of the universe constant.

Quantum fluctuations in the inflaton field when the universe was very small are believed to have magnified during the exponential expansion providing seedlike structures for the universe as a whole. These seeds led to the formation of the galaxies. This is an amazing connection between quantum theory and the large-scale structure of the universe.

Inflation theory makes several predictions that are consistent with observation to date.

The Kasner Metric

The *Kasner metric* is a solution to the Einstein field equations that has an interesting property that makes it useful from a string theory perspective. We can characterize the Kasner metric by considering the notion of *isotropy*. If space is isotropic, then it is the same in all directions. This is a reasonable assumption that is used routinely in cosmology when considering the 3 + 1 dimensional space-time we appear to live in. On large scales, it doesn't matter which direction you look—the universe looks the same.

In contrast, the Kasner metric is *anisotropic*, meaning that not all spatial dimensions evolve in the same way. As time increases, the universe expands in n of the spatial directions but contracts in the other $D - n$ directions. So this metric could describe a universe in which some of the dimensions become small (compactified) as the universe evolves. As you might imagine, this makes the metric appealing within the context of string theory.

The Kasner metric can be written in the following way:

$$ds^2 = -dt^2 + \sum_{j=1}^{D} t^{2p_j}(dx^j) \tag{16.6}$$

The presence of the term t^{2p_j} multiplying each spatial direction dx^j makes the behavior of each dimension dependent on the passage of time. We call the p_j *Kasner exponents* and they must satisfy two conditions aptly named the *Kasner conditions*:

$$\sum_{j=1}^{D-1} p_j = 1 \qquad \text{(first Kasner condition)}$$

$$\sum_{j=1}^{D-1} (p_j)^2 = 1 \qquad \text{(second Kasner condition)} \tag{16.7}$$

The Kasner conditions enforce a constraint on the p_j. What these tell us is that the p_j cannot all have the same sign. Since the metric term related to each spatial dimension depends on t^{2p_j}, this tells us that some dimensions will expand as time increases and some will contract as time increases. That is

- If p_j is positive, then $t^{2p_j} > 1$ and the direction x^j is increasing with time.
- If p_j is negative, then $t^{2p_j} < 1$ and the direction x^j is shrinking with time.

To see this, note a simple illustration. Let $p_j = 0.2$. Then at $t_1 = 5$, we have $t^{2p_j} = 5^{0.2} = 1.38$. At a later time $t_2 = 15$, we have $t^{2p_j} = 15^{0.2} = 1.72$, so the dimension has increased by a factor of $1.72/1.38 \approx 1.25$. Now suppose that instead $p_j = -0.2$. At $t_1 = 5$, we have $t^{2p_j} = 5^{-0.2} = 0.72$. At a later time $t_2 = 15$, we have $t^{2p_j} = 15^{-0.2} = 0.58$, so clearly the dimension is shrinking when the Kasner exponent is negative.

When the Kasner metric is studied in string theory, it must be supplemented by equations for the dilaton field ϕ. The dilaton field is related to the metric through the Kasner exponents p_j. In particular, it is possible to take

$$\phi = -\left(1 - \sum_{j=1}^{D} p_j\right)\ln t \tag{16.8}$$

Interestingly, the dilaton field introduces a type of duality into the model. In fact, this duality is related to T-duality, because it relates large and small distances. Given a set of Kasner exponents p_j and a dilaton field ϕ, there exists a dual solution with

$$p'_j = -p_j \qquad \phi' = \phi - 2\sum_{j=1}^{D} p_j \ln t \qquad (16.9)$$

Notice that since $p'_j = -p_j$, expanding dimensions in the theory are the contracting dimensions in the dual theory and vice versa.

Pre-big-bang cosmology can be described in terms of this duality. It allows for the universe to go through the following stages of evolution:

- It starts out in a large, flat, and cold state.

- It contracts to a self-dual point. The universe enters a state where it is small, highly curved, and very hot. This is the "big bang."

- It enters an expansion phase which is the universe we live in.

This was the first attempt at a cosmological model using string theory. However, it has since been discarded in favor of brane-based cosmological models. This is because several problems with the model could not be resolved, and brane models of the universe are compelling because of how the fields of the standard model and gravity are described. Before going on to brane-world cosmology though, let's see how the Kasner metric can describe an accelerating universe.

An interesting effect that can arise when considering some spatial dimensions contracting and others expanding is that the contracting dimensions actually cause the expanding dimensions to accelerate.[1] Suppose that we have $n > 1$ contracting dimensions with three expanding spatial dimensions. It can be shown that they cause the three spatial dimensions not only to expand, but to do so in an inflationary manner without a cosmological constant.

We write the number of space-time dimensions as $D = n + 4$, where we understand that the n dimensions which contract are all spatial and the remaining dimensions are $3 + 1$ dimensional space-time. The metric can be written in a general form which is split between time, the expanding dimensions, and the contracting dimensions as

$$ds^2 = -dt^2 + a^2(t)\left(\sum_{i=1}^{3} dx_i^2\right) + b^2(t)\left(\sum_{m=4}^{D-1} dx_m^2\right) \qquad (16.10)$$

---

[1] Levin, Janna, "Inflation from Extra Dimensions," *Phys. Lett.* vol. B343, 1995, 69–75.

Here, $a(t)$ is a scale factor associated with the three expanding spatial dimensions and $b(t)$ is a scale factor associated with the contracting spatial dimensions. Solving the Einstein equations in vacuum gives the following:

$$3\frac{\ddot{a}}{a} + n\frac{\ddot{b}}{b} = 0$$

$$\frac{\ddot{a}}{a} + 2(H_a + nH_b)H_a + 2\frac{k^{(3)}}{a^2} = 0$$

$$\frac{\ddot{b}}{b} + (3H_a + (n-1)H_b)H_b + \frac{n-1}{b^2}k^{(n)} = 0$$

Here we have introduced two Hubble constants. One is the usual Hubble constant associated with our expanding universe H_a:

$$H_a = \frac{\dot{a}}{a} \tag{16.11}$$

The second Hubble constant is associated with the contracting extra dimensions:

$$H_b = \frac{\dot{b}}{b} \tag{16.12}$$

The constants $k^{(3)}$ and $k^{(n)}$ are related to the local curvature and so can be $+1, 0, -1$. We choose the locally flat case and so set $k^{(3)} = k^{(n)} = 0$. This allows the equations to be simplified somewhat, giving three relations for the Hubble constants:

$$H_a^2 + nH_aH_b + \frac{n(n-1)}{6}H_b^2 = 0$$

$$\dot{H}_a + (3H_a + nH_b)H_a = 0$$

$$\dot{H}_b + (3H_a + nH_b)H_b = 0$$

The ratio H_b/H_a can then be written as

$$\frac{H_b}{H_a} = -\left(\frac{3n \pm \sqrt{3n^2 + 6n}}{n(n-1)}\right) \tag{16.13}$$

The choice of sign corresponds to an accelerating or decelerating universe (for the expanding extra dimensions). Of course, the choice of sign here is arbitrary, the model doesn't dictate why we would pick one sign or the other—it only describes that an accelerating universe is possible. We take the + sign for the accelerating case. Using $H_a^2 + nH_aH_b + [n(n-1)/6]H_b^2 = 0$, we can eliminate H_d from the equations and write an equation for H_a alone:

$$\frac{\dot{H}_a}{H_a} = \left(\frac{3+\sqrt{3n^2+6n}}{n-1}\right) \tag{16.14}$$

Now the accelerating nature of the expansion is apparent since $\dot{H}_a > 0$. Integration gives

$$-\frac{1}{H_a(t)} + \frac{1}{H_a(0)} = \left(\frac{3+\sqrt{3n^2+6n}}{n-1}\right)t$$

Now make the definition

$$\bar{t} = \frac{n-1}{3+\sqrt{3n^2+6n}}\frac{1}{H_a(0)}$$

Then it can be shown that the Hubble constant is given by

$$H_a(t) = \frac{H_a(0)}{1-t/\bar{t}} \tag{16.15}$$

Further integration gives the scale factor:

$$a(t) = \frac{\bar{a}}{(1-t/\bar{t})^p} \tag{16.16}$$

where $\bar{a}$ is a constant of integration and we have defined

$$p = \frac{-3+\sqrt{3n^2+6n}}{3(n+3)} \tag{16.17}$$

The acceleration of the three expanding dimensions of the universe is then

$$\frac{\ddot{a}}{a} = \left(\frac{p}{\bar{t}}\right)\frac{p+1}{\bar{t}}\frac{1}{(1-t/\bar{t})^2} > 0$$

Using the relation for the ratio H_b/H_a it can be shown that

$$b(t) = \bar{b}(1-t/\bar{t})^q \tag{16.18}$$

where $\bar{b}$ is a constant of integration and

$$q = \frac{n + \sqrt{3n^2 + 6n}}{n(n+3)} \tag{16.19}$$

This solution gives a Kasner type metric. Explicitly, we have

$$ds^2 = -dt^2 + \bar{a}\left(1 - \frac{t}{t}\right)^{-2p}\left(\sum_{i=1}^{3} dx_i^2\right) + \bar{b}\left(1 - \frac{t}{t}\right)^{2q}\left(\sum_{m=4}^{D-1} dx_m^2\right) \tag{16.20}$$

The Randall-Sundrum Model

The approaches described in the previous section are no longer considered tenable. The current line of research into cosmology from a string/M-theory perspective was launched with a brane-based approach called the *Randall-Sundrum model*.[2] This model is not a string/M-theory approach per se. Instead, it is simply a model which invokes the existence of extra dimensions and the existence of branes. Moreover, the model was not developed for the purposes of cosmology. The model was put forward as a possible solution to the *hierarchy problem* of particle physics. To review, the hierarchy problem is the fact that there is an enormous energy gap between the natural or fundamental energy scales of gravity and the electroweak theory. The electroweak scale is on the order of just 100 GeV, while the gravitational scale is on the order of a whopping 10^{18} GeV. The beauty of the Randall-Sundrum model is that it solves the hierarchy problem with a simple model based on branes and higher-dimensional space-time. We discuss the Randall-Sundrum model because the basic idea, two 3-branes connected along an extra spatial dimension, was the starting point for an idea of how to approach big-bang cosmology in string theory.

Now let's describe the basics of the model, which will form the basis of cosmological models more directly connected to string theory. We consider a five-dimensional space-time with two branes called the *visible brane* (our universe) and the *hidden brane*. The branes form boundaries to a five-dimensional region called *the bulk*. The branes have the usual 3 + 1 dimensional space-time. Gauge interactions are restricted to the brane, while gravity can propagate along the extra dimension and hence into the bulk, as well as in the branes.

We denote the extra spatial dimension by y and refer to the other space-time coordinates as x^μ. The five-dimensional metric is denoted by g_{AB}. The two branes

[2] First proposed by Lisa Randall and Raman Sundrum in "A Large Mass Hierarchy from a Small Extra Dimension", *Phys.Rev.Lett.* 83 (1999):3370–3373. Available on the arXiv at http://lanl.arxiv.org/abs/hep-ph/9905221.

have induced metrics given by $h^i_{\mu\nu} = g_{\mu\nu}(x^\mu, y_i)$, where $i = 1, 2$ for the visible and hidden branes, respectively. The Randall-Sundrum action is

$$S = \int dy\, d^4x \sqrt{-g}\left(\frac{M_5^3}{2}R - \Lambda\right) + \sum_{i=1}^{2} \int_i d^4x \sqrt{-h^{(i)}}\left(\Lambda_i + L^{(i)}_{\text{matter}}\right) \qquad (16.21)$$

The index i on the second integral indicates that we integrate over each brane separately. The additional terms included here are

- M_5: The Planck mass in five dimensions.
- Λ: The cosmological constant in the bulk.
- Λ_1 and Λ_2: The cosmological constants on the visible and hidden branes.
- R: The scalar curvature in five dimensions.
- $L^{(i)}_{\text{matter}}$: The lagrangian density for matter fields on the visible and hidden branes. On the visible brane, it is the standard model fields but could be different on the hidden brane.

The dimension y ranges over $0 \le y \le \pi r_c$, where r_c is a constant and the two branes are located at the boundaries. The visible brane is located at $y_1 = \pi r_c$, while the hidden brane is located at $y_2 = 0$.

Imposing a requirement that Poincaré invariance is respected, the following metric is chosen that is a slice of anti-de Sitter space:

$$ds^2 = e^{-2ky}\eta_{\mu\nu}dx^\mu dx^\nu + dy^2 \qquad (16.22)$$

The exponential term e^{-2ky} is called the *warp factor*. We will see that the warp factor connects mass scales in our $3 + 1$ dimensional universe to five-dimensional mass parameters.

It can be shown that the cosmological constants in the bulk and on each of the branes are given by

$$\Lambda = -6M_P^3 k^2$$
$$\Lambda_1 = -\Lambda_2 = -6M_P^3 k \qquad (16.23)$$

If $k < M_p$, this tells us that the space-time curvature of the bulk is small compared to the Planck scale.

The exponential warp factor causes the large gap between the observed Planck and electroweak scales. Moving to an effective four-dimensional theory, Randall and Sundrum showed that the Planck mass in four dimensions could be derived from the five-dimensional Planck mass via

$$M_P^2 = M_5^3 \int_{-\pi r_c}^{\pi r_c} dy\, e^{-2k|y|} = \frac{M_5^3}{k}\left(1 - e^{-2k\pi r_c}\right) \qquad (16.24)$$

A physical mass m on the visible three branes (our $3 + 1$ dimensional world) is related to a fundamental mass parameter m_0 in the underlying higher-dimensional theory by

$$m = m_0 e^{-k\pi r_c} \tag{16.25}$$

This allows us to obtain the electroweak scale where $m \sim 100$ GeV from the Planck mass $m_0 \sim 10^{18}$ GeV if $k\pi r_c \approx 37$. So the Randall-Sundrum model tells us that the scale of the electroweak interactions is a consequence of the curvature of space-time, as codified in the warp factor.

The Randall-Sundrum model has shed new light on the scales of particle physics, but other than setting up an arena with two branes and an extra dimension, it hasn't said anything about cosmology. But this setup sets the stage for an M-theory-based cosmology that allows the boundary branes to move along the extra dimension. We discuss this scenario in the next section.

Brane Worlds and the Ekpyrotic Universe

A cosmological model based on M-theory was proposed by Neil Turok and Paul Steinhardt.[3] In the Randall-Sundrum model, we have a five-dimensional universe with two branes fixed at the boundaries. Now imagine that instead the branes can move along the fifth dimension through the bulk. This idea is the origin of the *ekpyrotic universe,* a model fully rooted in string/M-theory. In particular, the ekpyrotic scenario is based on five-dimensional heterotic M-theory. The models are studied with five space-time dimensions because we start with 11 space-time dimensions in M-theory, and compactify six of the dimensions down to a tiny size which is irrelevant on cosmological scales.

In this model, we are imagining a universe which has always existed, but which goes through a cyclic pattern. This pattern begins with an initial state characterized by the boundary branes living in a flat, empty, and cold state. They are located at the boundaries of the fifth dimension and are parallel. As mentioned above, in the ekpyrotic scenario the branes are moving, so they move toward one another and collide. The collision of the branes, a process called *ekpyrosis* in the literature, is seen as the "big bang." The energy from the collision creates the matter in the brane. After collision, the branes move off apart from one another and cool down. Eventually they return to the cold, empty, flat initial state, and the process begins all over again. The driving force behind this is a scalar field ϕ called the *radion field,* which determines the distance between the branes. It causes the universe to evolve through a period of slow acceleration, followed by deceleration and contraction. It then triggers a bounce and reheating of the universe.

---

[3] See http://lanl.arxiv.org/abs/astro-ph/0204479 for an informal discussion.

The scenario depicted here solves many cosmological riddles, if it is to be believed. First let's consider two major riddles solved by inflation: homogeneity and isotropy. Inflation seeks to address this problem (explaining why the universe is homogeneous and isotropic) by postulating the existence of a field that turns on for a brief instant causing the universe to expand exponentially. While this scenario has been quantified in a plausible manner, it is not unreasonable to have doubts about a theory that describes a field that turns on for a flicker of an instant and turns off as fast, never to be seen again in the entire history of the universe. So what does the ekpyrotic scenario have to offer?

In the ekpyrotic scenario, there are two flat, parallel branes that collide like two nearly perfectly flat metal plates, say. Since the branes are parallel they collide at the same time (well almost anyway, let quantum theory intervene) at all points along the branes. This action endows the visible brane with the same energy density at all points with constant initial temperature called the *ekpyrotic temperature*. This explains why the universe looks the same everywhere in all directions and why the cosmic microwave background is the same everywhere—the universe began with the same initial conditions at all points.

The flatness problem is solved by setting the initial conditions of the branes to the vacuum state. In the vacuum state the branes are flat and empty, so no mysterious fine tuning of matter density is required to make the universe turn out flat. The reasonable assumption that the branes start off in the vacuum state forces them to be flat.

Now, of course, quantum theory means that everything is not as exact as described so far. Quantum fluctuations in the branes called *brane ripples* result from the movement of the branes along the fifth dimension. These fluctuations mean that not every point on the brane collides with the other brane at exactly the same instant. Instead, most will collide at some average time, while some will collide earlier than average and some will collide later than average. Hence, rather than producing a universe with an absolutely uniform temperature, the collision will produce a universe with some regions slightly colder than average (because they collided earlier) and some regions slightly hotter than average (because they collided later). These are the seeds the universe needs to produce the large scale structures of the universe like the galaxies. Once again, quantum effects are seen to give birth to large scale cosmological structure, providing a link between the very large and the very small in the universe.

One distasteful aspect of general relativity is the presence of "singularities" in the theory. These are points in space-time where quantities like curvature (the gravitational field) and temperature blow up to infinity. The "big-bang singularity" is one such example.

In the ekpyrotic model, the singularity is far milder than in classical general relativity. Two branes move toward each other, they collide, and then they bounce off and return to their initial positions. The "big bang" is an event that occurs with a large but *finite* temperature. There is no singularity corresponding to infinite curvature. Matter and radiation densities on the branes are finite. And there is no infinitely small point where all of matter, space, and time supposedly sprung from by magical fiat. However, there is singular behavior at the "big crunch" when the two branes collide, because the extra dimension between them disappears during the collision. After the branes separate and move off from each other the extra dimension reappears. Of course, while this model dispenses with much of the singular behavior of general relativity, it may be just as hard to believe that space and time always existed. In the end, experiment and observation will be our guides to determine in a scientific manner which scenario is closer to the truth.

The ekpyrotic scenario answers another mystery of cosmology, the origin of matter. During the collision the kinetic energy of motion of the branes is converted to heat or thermal energy. This is just like a car crash, where some of the energy of motion of the cars is converted to heat. In the case of the branes, the heat energy can be used to create matter via the Einstein relation $E = mc^2$.

The current form of the ekpyrotic scenario is called the *cyclic model* of the universe. It proposes that

- The big bang is not the origin of time.
- The universe always existed and runs through a repeated cycle of brane collisions.

A cycle in the history of the universe goes as follows:

- Two branes collide providing a big bang which acts as a transition between cycles. Matter and radiation are created.
- The hot big-bang phase creates large-scale structure in the universe.
- This is followed by a period of slow but accelerated expansion where the universe cools down and dilutes.

The ekpyrotic scenario provides an alternative to inflation that can be used to explain many cosmological mysteries. Suprisingly, they may be able to be distinguished by observational tests (at least in principle). Inflation predicts that gravitational waves are scale invariant. This is not the case for the ekpyrotic model.

Summary

We began exploring cosmological scenarios by considering the Kasner metric, which allows some dimensions to contract while others expand as the universe evolves. Models of this type are not satisfactory and so have been discarded. The Randall-Sundrum model imagines the universe to be constructed out of two branes that bound a higher-dimensional bulk. This idea was extended in the ekpyrotic model, which allows the branes to move and collide, explaining the big bang and providing a string theory alternative to inflation. The ekpyrotic scenario does not say the big bang never happened, rather it explains the big bang without evoking a singularity. Once the brane collision has occurred, the universe evolves according to standard big-bang theory on the branes.

Quiz

1. Let $g = \det g_{\mu\nu}$, wherge $g_{\mu\nu}$ is the Kasner metric. Using the first Kasner condition, find an expression for $\sqrt{-g}$.

2. Suppose that $p_j = \dfrac{1}{D-1}$ for all j in the Kasner metric. Is the second Kasner condition satisfied?

3. Consider the metric derived in the text:

$$ds^2 = -dt^2 + \bar{a}\left(1 - \frac{t}{\bar{t}}\right)^{-2p}\left(\sum_{i=1}^{3} dx_i^2\right) + \bar{b}\left(1 - \frac{t}{\bar{t}}\right)^{2q}\left(\sum_{m=4}^{D-1} dx_m^2\right)$$

Show that the Kasner conditions are satisfied.

Final Exam

1. Consider the lagrangian for a classical string $L = \dfrac{1}{2\pi\alpha'}\sqrt{\dot{X}_\mu^2 X'^{\mu 2} - (\dot{X}_\mu X'^\mu)^2}$.
   Write down the canonical momentum.

2. Consider a classical string which is in a configuration described by a rigid
   rod rotating about the origin with angular velocity ω. Find the energy of the
   string from $E = \dfrac{1}{2\pi\alpha'}\displaystyle\int_{-l}^{l} d\sigma \,\dfrac{1}{\sqrt{1 - \omega^2\sigma^2}}$.

3. Consider the action for a p-brane with cosmological constant
   $S = -\dfrac{T}{2}\displaystyle\int d^{p+1}\sigma\sqrt{-h}\, h^{\alpha\beta}\partial_\alpha X \cdot \partial_\beta X + \Lambda\int d^{p+1}\sigma\sqrt{-h}$. Find the classical equations
   of motion for the metric.

4. Using the action of the previous problem, find a constraint on the
   cosmological constant Λ using $h^{\mu\nu}h_{\mu\nu} = p + 1$.

5. Consider the classical string, describing its dynamics using the Polyakov
   action. What form does the action take if the worldsheet metric is taken to
   be flat?

6. Using the action of Prob. 5, what is the canonical momentum?

7. The mass of a classical open string is $M^2 = \dfrac{1}{\alpha'}\sum_{n=1}^{\infty}\alpha_{-n}\cdot\alpha_n$. How is the mass of a closed string different?

8. Consider the classical string. What is the algebra satisfied by the Virasoro generators?

9. Using the normal ordering prescription, $L_m = \dfrac{1}{2}\sum_{n=-\infty}^{\infty} :\alpha_{m-n}\cdot\alpha_n:$ find L_0.

10. What is the mass-shell condition for states of the bosonic string, written in terms of Virasoro operators?

In Probs. 11–14, consider the bosonic string.

11. Find the angular momentum operators $J^{\mu\nu}$.

12. Using the result of Prob. 11, find $\left[p_0^{\mu}, J^{\rho\lambda}\right]$.

13. Find $[J^{\mu\nu}, J^{\rho\lambda}]$.

14. For the Virasoro operator L_m, find $[L_m, J^{\mu\nu}]$.

15. Consider the first excited state of the open bosonic string. Let ξ_{μ} be a polarization vector and consider the action of L_1 on the state $\xi\cdot\alpha_{-1}|0,k\rangle$. What condition on the polarization and momentum follows from the Virasoro constraint $L_1|\psi\rangle = 0$ for physical states $|\psi\rangle$?

16. What condition cancels the conformal anomaly for the bosonic string?

17. Consider the Polyakov action in the conformal gauge. State the constraints on the components of the energy momentum tensor.

18. State the Neumann boundary conditions for classical, open bosonic strings.

19. Consider a relativistic point particle with space-time coordinates $x^{\mu}(\tau)$ and action $S = -m\int d\tau\sqrt{-\dot{x}^{\mu}\dot{x}_{\mu}}$, where $\dot{x}^{\mu} = \dfrac{dx^{\mu}}{d\tau}$. Use the usual variational procedure to find the equations of motion, and write down these equations in terms of the conjugate momentum.

20. Consider your solution to Prob. 19. What is the condition on $\dot{p}^{\mu}$?

21. Consider your solution to Prob. 19. Take the *static gauge*, where $x^0 = t$. What is the action in this case if we use the usual definition of the particle velocity $\vec{v} = \dfrac{d\vec{x}}{dt}$?

22. What is the momentum in this case (using the action from Prob. 21)?

23. How do the spinors ψ_{μ} in the RNS formalism transform under Lorentz transformations?

In problems 24–26, let Γ^{μ} be a gamma matrix in $D = 10$ space-time dimensions and following the text let $\Gamma_{11} = \Gamma_0\Gamma_1\cdots\Gamma_9$.

24. Calculate $\{\Gamma_{11}, \Gamma^{\mu}\}$.

25. Calculate $\Gamma_{11}\Gamma^0$.

26. Find a simple expression for $\Gamma^0 \Gamma_\mu \Gamma^0$.

27. What is the unifying principle of supersymmetry?

28. What characterizes a supersymmetry transformation?

29. In string theory, there are two general approaches to introducing supersymmetry to the theory. What are they?

30. What are the supersymmetry transformations in the RNS formalism?

In problems 31–33, consider the uppercurrent in the RNS formalism

$$J_a = \frac{1}{2} \rho^b \rho_a \psi^\mu \partial_b X_\mu, \text{ where } a, b = 0, 1.$$

31. Calculate $\partial_a J^a$.

32. Write down "ladder operator" type expressions $J_\pm = \frac{1}{2}(J_0 \pm J_1)$.

33. Find equations satisfied by $J_\pm = \frac{1}{2}(J_0 \pm J_1)$.

34. What are the Ramond boundary conditions?

35. What are the Neveu-Schwarz boundary conditions?

36. What type of space-time states arise from Ramond boundary conditions?

37. What type of space-time states arise from the NS sector?

38. What is the Majorana condition?

39. How does a Majorana-Weyl spinor differ from a general Dirac spinor?

40. How are the modal expansions for the R sector and NS sector different?

41. Consider a closed string. If the boundary conditions for left movers and right movers are NS and R, respectively, what type of space-time state is described?

42. Consider a closed string. If the boundary conditions for left movers and right movers are both NS, what type of space-time state is described?

43. Consider a closed string. If the boundary conditions for left movers and right movers are both R, what type of space-time state is described?

44. How are the Virasoro operators L_m of bosonic string theory generalized in the RNS formalism?

45. Consider the operator $F_m = \sum_n \alpha_{-n} \cdot d_{m+n}$ associated with the R sector. Perform an explicit calculation to determine the anticommutator $\{F_m, F_n\}$ in 10 space-time dimensions.

46. Adding supersymmetry to string theory eliminates a lot of particle states. In particular, which particle state does it remove which makes bosonic theory unstable?

47. Using the action $S = -\frac{T}{2} \int d^2\sigma \, (\partial_\alpha X^\mu \partial^\alpha X_\mu - i \bar{\psi}^\mu \rho^\alpha \partial_\alpha \psi_\mu)$, follow the usual variation procedure to deduce the equations of motion obeyed by ψ_0^μ and ψ_1^μ.

48. Consider the RNS formalism. What condition on the energy-momentum tensor would indicate that negative-norm states have been removed from the theory?

49. Calculate $[p_0, p_1]$.

50. Let a and b be two Grassman numbers. What is their defining characteristic?

51. Why is the Dirac delta function $\delta(\sigma - \sigma')$ included in the commutation relations of string theory?

52. In conformal field theory, we define the coordinates of the string (τ, σ) in terms of a complex variable z and its complex conjugate. How can this be done?

53. If you are given that $\langle X(w)X(z)\rangle \sim \log(w - z)$, find $\langle e^{ikX(w)}e^{-ikX(z)}\rangle$.

54. A field $\phi(z)$ has conformal weight h and so under $z \rightarrow z_1(z)$ transforms as

$$\phi(z) = \tilde{\phi}(z_1)\left(\frac{dz_1}{dz}\right)^h.$$ How does it transform under a second transformation

$z_1(z) \rightarrow z_2[z_1(z)]$?

55. By examining its behavior under a conformal transformation, find the conformal weight of the bosonic field X^μ.

56. Find the conformal weight of ∂X.

57. Find the conformal weight of $\partial X \cdot \partial X$.

58. Calculate $T(z)T(w)$.

59. What can you conclude for your result in Prob. 58?

60. Let θ be a Grassman variable and define the supersymmetry derivative

$$\frac{\partial}{\partial \theta} + \theta\frac{\partial}{\partial z}.$$ Find D^2.

61. Do the Majorana-Weyl fermions ψ_-^μ used in the RNS formalism describe left movers or right movers?

62. What is the normal ordering constant for the R sector?

63. What is the normal ordering constant for the NS sector?

64. Given the winding number n, what is the winding w?

65. Consider bosonic string theory. If the 25th dimension is compactified, what is the winding mode?

66. How is the level matching condition modified if a single spatial dimension is compactified on a circle of radius R?

67. If a single spatial dimension is compactified on a circle of radius R, what is the mass formula for a closed string?

68. When a single spatial dimension is compactified on a circle of radius R, there are extra terms in the expression used to determine the mass of the state. What factors are these? Is the mass increased or decreased?

69. What is the effect of compactification on the center of mass momentum?

70. Consider the compactification of a single spatial dimension to a circle of radius R and consider the limiting behavior as $R \to 0$. Why do the winding states go to the continuum limit?

71. How does T-duality relate winding and momentum states?

72. How does T-duality relate distance scales in different theories?

73. What string theories are related by T-duality?

74. A certain string theory has orientable strings. What does this mean?

75. What is the mass of a tachyon state in bosonic string theory?

76. How is Type I string theory different from all the other superstring theories?

77. What symmetry groups are associated with heterotic string theory?

78. How does the Kasner metric describe the behavior of space-time as the universe evoles?

79. What is the effect of the dilaton field on the Kasner metric?

80. What is the self-dual point?

81. In the Horava-Witten model, how are the uncompactified dimensions laid out?

82. What condition on the beta function of the dilaton field must be satisfied for scale invariance?

83. What property of space-time is implied by conformal invariance as related to the dilaton field?

84. How is the vanishing of the beta function related to general relativity?

85. Einstein's relativity allows for the description of "extreme" black holes that carry electric charge. How is this extended in string theory?

86. Let K be the Kaluza-Klein excitation and n the winding number. How is a state (K,n) in a theory with a compactified dimension of radius R transformed under T-duality?

87. How are the number operators transformed under T-duality?

88. A state has mass m. How is the mass changed under T-duality?

89. How are the equations of motion related for the compactified dimension under T-duality between a theory and its dual?

90. Consider an open bosonic string under a T-duality transformation. What happens to the momentum of the dual string along the compactified dimension?

91. The string momenta along a certain spatial direction are found to be fractional. How are the fields described in the dual theory?

92. What type of boundary condition is $\partial_\sigma X^\mu\big|_{\sigma=0,\pi}=0$?

93. In type II A superstring theory, what dimensions are allowed for Dp-branes?

94. If the value of the fields X^μ are specified on the boundary, what type of problem is being posed?

In Probs. 95–96 let $D=10$ and suppose that there is a pure electric background field $F_{0i}=E_i$.

95. What is the form of the Born-Infeld action?

96. What is the maximum value of the electric field?

97. How is the dilaton field ϕ related to string coupling g_s?

The final problems will be very challenging for many readers. For problems 98, 99, and 100, suppose that you have closed strings with periodic boundary conditions for $\mu=0,1,...,24$ but $X^{25}(\sigma,\tau)=-X^{25}(\sigma+\ell_s,\tau)$. Use the light-cone gauge (based on Polchinski 1.9).

98. How does the modal expansion change for X^{25} as opposed to the usual modal expansion for closed bosonic strings?

99. Using $\Pi^\mu=\dfrac{p^+}{\ell_s}\partial_\tau X^\mu$ and $\displaystyle\sum_{n=1}^{\infty}(n-\theta)=\dfrac{1}{24}-\dfrac{1}{8}(2\theta-1)^2$ find the hamiltonian

from $H=\dfrac{\ell_s}{4\pi\alpha' p^+}\displaystyle\int_0^{\ell_s} d\sigma\left(2\pi\alpha'\Pi^i\Pi^i+\dfrac{1}{2\pi\alpha'}\partial_\sigma X^i\partial_\sigma X^i\right).$

100. What is the mass spectrum? Note that the level-matching condition is still satisfied.

Quiz Solutions

Chapter 1

1. b		6. a	
2. a		7. c	
3. c		8. b	
4. b		9. c	
5. d		10. a	

Chapter 2

1. Use $\dfrac{d}{d\tau}\left(\dfrac{\partial L}{\partial \dot{a}}\right)=\dfrac{\partial L}{\partial a}$.

2. $\dfrac{\partial^2 X_\mu}{\partial \tau^2}-\dfrac{\partial^2 X_\mu}{\partial \sigma^2}=0$

3. The Polyakov action is invariant under a Weyl transformation.

4. $h_{\alpha\beta} = \partial_\alpha X^\mu \partial_\beta X^\nu \eta_{\mu\nu}$

5. $X_{cm}^\mu = x^\mu + 2\ell_s^2 p^\mu \tau$

6. $\dfrac{1}{\sqrt{2}\ell_s}\alpha_0^\mu$

Chapter 3

1. $\delta h^{\alpha\beta} = \partial^\alpha \varepsilon^\beta + \partial^\beta \varepsilon^\alpha - \partial_\rho h^{\alpha\beta}\varepsilon^\rho$

2. The derivative can be evaluated as follows:

$$\frac{dp_\mu}{d\tau} = \int_0^{\sigma_1} d\sigma \frac{dP_\mu^\tau}{d\tau} = -\int_0^{\sigma_1} d\sigma \frac{dP_\mu^\tau}{d\sigma} = P_\mu^\tau(\sigma = 0) - P_\mu^\tau(\sigma = \sigma_1)$$

3. Consider $\dfrac{\delta S_p}{\delta\phi} = \dfrac{\delta S_p}{\delta h^{\alpha\beta}} \dfrac{\delta h^{\alpha\beta}}{\delta\phi}$.

4. Use $T_{++} = T_{--} = 0$.

5. $\partial^\tau J_\tau^{\mu\nu} + \partial^\sigma J_\sigma^{\mu\nu} = 0, J^{\mu\nu} = \int d\sigma J_\tau^{\mu\nu} - \int d\tau J_\sigma^{\mu\nu}$, boosts and rotations

Chapter 4

1. $0,0$

2. $i\eta^{\mu\nu}$

3. 0

4. $\dfrac{1}{\alpha'}(1-a)$

5. $\left[\alpha_m^i, \alpha_n^j\right] = m\delta^{ij}\delta_{m+n,0}$

6. $\left[\bar{\alpha}_m^i, \bar{\alpha}_n^j\right] = m\delta^{ij}\delta_{m+n,0}, \left[\alpha_m^i, \bar{\alpha}_n^j\right] = 0$

Chapter 5

1. 0

2. No, $T_b\left(T_a(z)\right) = \dfrac{1+az}{1+az+b} \neq T_{a+b}(z)$

3. $i(\ell_1 - \overline{\ell}_1)$

4. $\dfrac{1}{4}\langle 0|x^\mu x^\nu|0\rangle + \dfrac{\eta^{\mu\nu}\ell_s^2}{4}\ln\overline{z} - \dfrac{\eta^{\mu\nu}\ell_s^2}{2}\ln(\overline{z}-\overline{z}')$

5. $\dfrac{\eta^{\mu\nu}\ell_s^2}{2}\dfrac{1}{(\overline{z}-\overline{z}')^2}$

6. $\langle 0|x^\mu x^\nu|0\rangle$

7. $\langle 0|x^\mu x^\nu|0\rangle - \dfrac{\eta^{\mu\nu}}{2\pi T}(\ln z + \ln\overline{z})$

$$-\dfrac{\eta^{\mu\nu}}{4\pi T}\left\{\ln\left(1-\dfrac{z'}{z}\right) + \ln\left(1-\dfrac{z'}{\overline{z}}\right) + \ln\left(1-\dfrac{\overline{z}'}{z}\right) + \ln\left(1+\dfrac{\overline{z}'}{\overline{z}}\right)\right\}$$

8. $-\dfrac{\eta^{\mu\nu}}{4\pi T}\left\{\ln(z-z')(z-\overline{z}') + \ln(\overline{z}-\overline{z}')(\overline{z}-z')\right\}$

9. $-\ell_s^2\left(\dfrac{\ell_s^2 k^2}{2(z-w)^2}e^{ik\cdot X(w)} + \dfrac{1}{z-w}\partial_w e^{ik\cdot X(w)}\right)$

Chapter 6

1. 0

2. $\oint\left(\dfrac{3}{2}\partial^2 c\partial c + \left(\dfrac{D}{12}-\dfrac{2}{3}\right)\partial^3 cc\right)$

3. 26

4. $k^2 = \dfrac{1}{\alpha'}$

Chapter 7

1. You will need the supplementary boundary condition
$$\int_0^{2\pi} d\sigma \int_{-\infty}^{\infty} d\tau\left[\partial_+(\psi_-\delta\psi_-) + \partial_-(\psi_+\delta\psi_+)\right] = 0$$

2. $i\dfrac{T}{2}\overline{\psi}^\mu\rho_\alpha\psi^\nu$.

3. $J_\alpha^{\mu\nu} = \dfrac{T}{2}(X^\mu \partial_\alpha X^\nu - X^\nu \partial_\alpha X^\mu + i\bar{\psi}^\mu \rho_\alpha \psi^\nu)$

4. 0

5. $\delta L_B = \partial_\beta \varepsilon^\alpha \left(\partial^\beta X_\mu \partial_\beta X^\mu - \dfrac{1}{2}\delta_\alpha^\beta \partial^\lambda X_\mu \partial_\lambda X^\mu \right)$

6. 0

7. 0

8. 0

9. 1

10. Massless vector boson, spin-3/2 fermion

Chapter 8

1. $\dfrac{\alpha'}{4}\left(p_R^{25}\right)^2 + N_R = \dfrac{\alpha'}{4}\left(p_L^{25}\right)^2 + N_L$

Chapter 9

1. Use $\left[\bar{\varepsilon}Q, \theta_A\right] = \bar{\varepsilon}_B \dfrac{\partial \theta_A}{\partial \bar{\theta}^B} + i\left[\bar{\varepsilon}_B \rho_{BC}^\alpha \theta_C, \theta_A\right]\partial_\alpha$

2. 0

3. Γ_μ

4. 0

Chapter 10

1. c

2. b

3. a

4. c

5. b

6. d

Chapter 11

1. c
2. b
3. b
4. a
5. c

Chapter 12

1. There are eight states, because there are eight transverse directions in space-time in the light-cone gauge.

2. These are the 16 states associated with the extra bosonic dimensions.

3. This keeps them in the left-moving sector.

4. $\left[\alpha_m^i, \alpha_n^j\right] = \left[\tilde{\alpha}_m^i, \tilde{\alpha}_n^j\right] = m\delta_{m+n,0}\delta^{ij}$ $\left[\tilde{\alpha}_m^I, \tilde{\alpha}_n^J\right] = m\delta_{m+n,0}\delta^{IJ}$

5. 0

Chapter 13

1. $\phi = 0, \pm 1$

2. $\phi = 0, m^2 = -\dfrac{A}{3}$ (tachyon), $\phi = \pm 1, m^2 = \dfrac{4A}{3} > 0$ (not a tachyon)

3. $\phi = 0, \pm\phi_0$

4. $\phi = 0, m^2 = -\dfrac{\phi_0}{2\alpha'}$

5. The stretching adds an energy $\dfrac{1}{2\alpha'}\alpha_0^a\alpha_0^a$ to the string where

$\dfrac{1}{2\alpha'}\alpha_0^a\alpha_0^a = \left(\dfrac{\bar{x}_2^a - \bar{x}_1^a}{2\pi\alpha'}\right)^2$ comes from the stretching between the branes.

Chapter 14

1. $T = \dfrac{\sqrt{(mG_4)^2 - Q^2 G_4}}{2\pi r_+^2}$

2. $T \le T_H = \dfrac{1}{4\pi\sqrt{\alpha'}}$

3. $\sim 10^{-8}$ K

4. 2×10^{68} years

5. $S = 2\pi\sqrt{Q_1 Q_5 n - J^2}$

Chapter 15

1. $F \approx \dfrac{z^2}{\sqrt{ag_s N}}$

2. It transforms the metric to the anti-de Sitter form

$$ds^2 \approx R^2\left[z^2(dt^2 - dx^2) - \frac{1}{z^2}dz^2 \right].$$

3. b

4. c

5. b

Chapter 16

1. $\sqrt{-g} = t^{p_1 + p_2 + \cdots + p_{D-1}} = t$

2. No, because $\displaystyle\sum_{j=1}^{D-1} p_j^2 = \dfrac{1}{D-1} \ne 1$. If $p_j = \dfrac{1}{D-1}$ for all j, then this is an isotropic universe. This shows that the Kasner metric cannot describe an isotropic universe if Kasner conditions are applied.

3. It is necessary to incorporate the fact that we are applying p to all three expanding dimensions and q to all n contracting dimensions. So the Kasner conditions are $-3p + nq = 1$, $3p^2 + nq^2 = 1$.

Final Exam Solutions

1. $P_\mu = \dfrac{1}{2\pi\alpha'} \dfrac{(X')^2 \dot{X}_\mu - (\dot{X}_\nu X'^\nu) X'_\mu}{\sqrt{\det |\partial_\alpha X^\nu \partial_\beta X_\nu|}}$

2. $E \sim \dfrac{l}{2\alpha'}$

3. $T\left\{\partial_\mu X \cdot \partial_\nu X - \dfrac{1}{2} h_{\mu\nu} (h^{\alpha\beta} \partial_\alpha X \cdot \partial_\beta X)\right\} + \Lambda h_{\mu\nu} = 0$

4. $\Lambda = \dfrac{T}{2}(p-1)$

5. $S = \dfrac{T}{2} \displaystyle\int d^2\sigma (\dot{X}^2 - X'^2)$

6. $P^\mu = T\dot{X}^\mu$

7. The mass of a closed string must include left movers and right movers. Hence it is given by $M^2 = \dfrac{2}{\alpha'} \sum\limits_{n=1}^{\infty} (\alpha_{-n} \cdot \alpha_n + \bar{\alpha}_{-n} \cdot \bar{\alpha}_n)$.

8. $[L_m, L_n] = i(m-n)L_{m+n}$

9. $L_0 = \dfrac{1}{2}\alpha_0^2 + \sum\limits_{n=1}^{\infty} \alpha_{-n} \cdot \alpha_n$

10. $(L_0 - 1)|\psi\rangle = 0$

11. $J^{\mu\nu} = x_0^{\mu} p_0^{\nu} - x_0^{\nu} p_0^{\mu} - i\sum\limits_{n=1}^{\infty} \dfrac{1}{n}\left(\alpha_{-n}^{\mu}\alpha_n^{\nu} - \alpha_{-n}^{\nu}\alpha_n^{\mu}\right)$

12. $[p_0^{\mu}, J^{\rho\lambda}] = -i\eta^{\mu\rho}p_0^{\lambda} + i\eta^{\mu\lambda}p_0^{\rho}$

13. $[J^{\mu\nu}, J^{\rho\lambda}] = i\eta^{\mu\rho}J^{\nu\lambda} + i\eta^{\nu\lambda}J^{\mu\rho} - i\eta^{\nu\rho}J^{\mu\lambda} - i\eta^{\mu\lambda}J^{\nu\rho}$

14. $[L_m, J^{\mu\nu}] = 0$

15. $\xi \cdot k = 0$

16. Setting the number of space-time dimensions to $D = 26$

17. $T_{00} = T_{11} = \dfrac{1}{2}(\dot{X}^2 + X'^2) = 0,\; T_{01} = T_{10} = \dot{X} \cdot X' = 0$

18. $\dfrac{\partial X^{\mu}}{\partial \sigma} = 0$ at the string endpoints.

19. $p^{\mu} = \dfrac{m\dot{x}^{\mu}}{\sqrt{-\dot{x}^{\nu}\dot{x}_{\nu}}}$

20. Momentum is constant, so $\dot{p}^{\mu} = 0$.

21. $S = -m\int dt\sqrt{1-v^2}$

22. $\vec{p} = \dfrac{m\vec{v}}{\sqrt{1-v^2}}$

23. As space-time vectors

24. 0

25. $-\Gamma^0\Gamma_{11}$

26. $\Gamma_{\mu}^{\dagger}$

27. Supersymmetry is a symmetry that unifies fermions and bosons.

28. It takes fermions into bosons and vice versa.

29. RNS formalism uses worldsheet supersymmetry, while GS formalism uses space-time supersymmetry.

30. $$\delta X^\mu = \bar{\varepsilon}\psi^\mu$$
    $$\delta\psi^\mu = -i\rho^\alpha\partial_\alpha X^\mu \varepsilon$$

31. 0

32. $J_+ = \dfrac{1}{2}\psi_1^\mu(\partial_\tau + \partial_\sigma)X_\mu \qquad J_- = \dfrac{1}{2}\psi_0^\mu(\partial_\tau - \partial_\sigma)X_\mu$

33. $(\partial_\tau - \partial_\sigma)J_+ = (\partial_\tau + \partial_\sigma)J_- = 0$

34. $\psi_\mu^+(\pi,\tau) = \psi_\mu^-(\pi,\tau)$, periodic boundary condition.

35. e $\psi_\mu^+(\pi,\tau) = -\psi_\mu^-(\pi,\tau)$, antiperiodic.

36. Fermions

37. Bosons

38. The components of a spinor are real.

39. They have real components, and have half the number of overall components.

40. In the R sector, the summation is over integers, in the NS sector it is over half integers.

41. Fermion

42. Boson

43. Boson

44. A fermionic operator is added, as in $L_m \rightarrow L_m^{(B)} + L_m^{(F)}$.

45. $\{F_m, F_n\} = 2L_{m+n} + 5m^2\delta_{m+n,0}$

46. The tachyon

47. $(\partial_\tau + \partial_\sigma)\psi_0^\mu = (\partial_\tau - \partial_\sigma)\psi_1^\mu = 0$

48. $\langle T_{\alpha\beta}\rangle = 0$

49. $2\rho_3$

50. They anticommute, that is, $ab + ba = 0$.

51. To ensure that operators do commute at different points σ along the string.

52. $\tau = \dfrac{z + \bar{z}}{2}$ $\qquad \sigma = \dfrac{z - \bar{z}}{2i}$

53. $(w - z)^{-k^2}$

54. $\phi \to \phi(z_2) \left(\dfrac{dz_2}{dz_1} \right)^h \left(\dfrac{dz_1}{dz} \right)^h$

55. 0

56. 1

57. 2

58. $T(z)T(w) \propto \dfrac{D - 26}{(z - w)^4} + \cdots$

59. $D = 26$

60. $\dfrac{\partial}{\partial z}$

61. Right movers

62. 0

63. 1/2

64. $w = \dfrac{nR}{\alpha'}$

65. $\dfrac{1}{2} \left(p_L^{25} - p_R^{25} \right) = nR$

66. $N_R - N_L = nK$

67. $\alpha' m^2 = \left(\dfrac{nR}{\alpha'} \right)^2 + \left(\dfrac{K}{R} \right)^2 + 2(N_R + N_L) - 4$

68. Kaluza-Klein excitations and winding increase the rest-energy of the string. The mass is increased.

69. It becomes quantized.

70. It costs less energy to wrap around the small extra dimension.

71. The winding states in one theory become the Kaluza-Klein excitations in the dual theory, and vice versa.

72. T-duality relates a small compactified dimension in one theory to a large dimension in the dual theory.

73. Types II A and II B, and the two heterotic theories.

74. You can tell direction along the string.

75. $m^2 = -1/\alpha'$

76. It contains open and closed strings, all other superstring theories describe closed strings.

77. There are two theories, one with $E_8 \times E_8$ and one with $SO(32)$.

78. The universe expands in n directions but contracts in D-n directions as time increases.

79. It makes contracting and expanding solutions dual to one another.

80. It defines a minimum radius such that $R_{min} = \sqrt{\alpha'}$.

81. There are two 3-branes separated along a fourth spatial dimension.

82. $\beta^\phi \to 0$

83. The number of space-time dimensions is 26.

84. It is equivalent to Einstein's equation for a scalar field.

85. Extreme black holes in string theory can have magnetic charge as well.

86. In a theory with $R' = \alpha'/R$, the state is transformed to (n, K).

87. They are unchanged.

88. The state has the same mass m in the dual theory.

89. $\partial_+ \tilde{X}_{25} = \partial_+ X_{25}$ $\partial_- \tilde{X}_{25} = -\partial_- X_{25}$

90. It vanishes.

91. The winding numbers are fractional.

92. Neumann

93. $p = 0, 2, 4, 6, 8$

94. Dirichlet

95. $S = \dfrac{1}{g_s} \left(\dfrac{T}{2\pi} \right)^5 \int d^{10} x \sqrt{1 - \left(\vec{E}/T \right)^2}$

96. $E = T = \dfrac{1}{2\pi\alpha'}$

97. $e^\phi = \dfrac{1}{g_s}$

98. Constant terms $x^{25} = p^{25} = 0$, sum over modes goes to sum over half-integral modes, that is, $\dfrac{\alpha_n^{\mu}}{n} \to \dfrac{\alpha_{n+1/2}^{\mu}}{n+1/2}, \dfrac{\tilde{\alpha}_n^{\mu}}{n} \to \dfrac{\tilde{\alpha}_{n+1/2}^{\mu}}{n+1/2}$.

99. $H = \dfrac{p^i p^i}{2p^+} + \dfrac{1}{\alpha' p^+} \sum_{n=1}^{\infty} \left[N + \tilde{N} - \dfrac{D-3}{12} + \dfrac{1}{24} \right]$

100. $\alpha' m^2 = 4\left(N - \dfrac{15}{16} \right)$

References

Books

Becker, K., M. Becker, and J. Schwarz, *String Theory and M-Theory: A Modern Introduction*, Cambridge University Press, New York, 2007.

Green, M., J. Schwarz, and E. Witten, *Superstring Theory,* vol. 1: *Introduction (Cambridge Monographs on Mathematical Physics)*, Cambridge University Press, New York, 1988.

Kaku, M., *Introduction to Superstrings and M-Theory*, Springer-Verlag, New York, 1999.

Kaku, M., *Strings, Conformal Fields, and M-Theory*, Springer-Verlag, New York, 2000.

Kiritsis, E., *String Theory in a Nutshell*, Princeton University Press, Princeton, N.J., 2007.

Maggiore, M., *A Modern Introduction to Quantum Field Theory*, Oxford University Press, New York, 2005.

Polchinski, J., *String Theory,* vol. 1: *An Introduction to the Bosonic String (Cambridge Monographs on Mathematical Physics)*, Cambridge University Press, New York, 1998.

Susskind, L., and J. Lindesay, *An Introduction to Black Holes, Information and the String Theory Revolution: The Holographic Universe*, World Scientific Publishing Company, Singapore, 2005.

Szabo, R., *An Introduction to String Theory and D-Brane Dynamics*, Imperial College Press, London, 2004.

Zwiebach, B., *A First Course in String Theory*, Cambridge University Press, New York, 2004.

Papers

Alvarez, E., L. Alvarez-Guame, and Y. Lozano, "An Introduction to T-Duality in String Theory," lanl.arXiv.org (1994), arXiv:hep-th/9410237v2. Also available in *Nucl. Phys. Proc. Suppl.* 41 (1995) 1–20.

Clifford, V. J., "D-Brane Primer," http://www.Citebase.org/abstract?id=oai%AarXiv.org%3Ahep-th%2F0007170.

Kiritsis, E., "Introduction to Non-perturbative String Theory," lanl.arXiv.org (1997), arXiv:hep-th/9708130v1.

Mohaupt, T., "Introduction to String Theory," lanl.arXiv.org (2002), arXiv:hep-th/0207249v1. Also available in *Lect. Notes Phys.* 631 (2003) 173–251.

Ooguri, H., "Gauge Theory and String Theory: An Introduction to the Ads/CFT Correspondence," lanl.arXiv.org (1999), arXiv:hep-lat/9911027v1. Also available in *Nucl. Phys. Proc. Suppl.* 83 (2000) 77–81.

Susskind L., and E. Witten, "The Holographic Bound in Anti–de Sitter Space," http://xxx.lanl.gove/abs/hep-th/9805114; and Maldacena, J. M., "The Large N Limit of Superconformal Field Theories and Super Gravity," *Adv. Theor. Math. Phys.* vol. 2, 1998, 231–252.

t'Hooft, G., "String Theory Lectures," online at www.phys.uu.nl/~thooft/lectures/string.html.

INDEX